Computers in the Laboratory
Current Practice and Future Trends

ACS SYMPOSIUM SERIES **265**

Computers in the Laboratory
Current Practice and Future Trends

Joseph G. Liscouski, EDITOR
Digital Equipment Corporation

Based on a symposium sponsored by
the Division of Computers in Chemistry
at the 186th Meeting
of the American Chemical Society,
Washington, D.C.,
August 28–September 2, 1983

American Chemical Society, Washington, D.C. 1984

Library of Congress Cataloging-in-Publication Data

Computers in the laboratory.
 (ACS symposium series, ISSN 0097-6156; 265)

 "Developed from a symposium sponsored by the
Division of Computers in Chemistry at the 186th
Meeting of the American Chemical Society,
Washington, D.C., August 28-September 2, 1983."

 Bibliography: p.
 Includes index.

 1. Chemical laboratories—Data processing—
Congresses. 2. Chemical laboratories—Automation—
Congresses. 3. Chemistry, Analytic—Data processing—
Congresses.

 I. Liscouski, Joseph G., 1945-   . II. American
Chemical Society. Division of Computers in Chemistry.
III. Series.

QD51.C65   1984   542'.028'54   84-18518
ISBN 0-8412-0867-0

FOREWORD

CONTENTS

PREFACE

A LABORATORY THAT TOOK ADVANTAGE of all of the forms of automation available today would be quite a place. Many routine sample preparation tasks would be done by robots that would introduce the samples to instruments for analysis. The data generated would be taken by a computer system, analyzed, reported, and then stored for later retrieval and more detailed analysis.

Each chemist, technician, secretary, and manager might have his own work station at his desk with communication between fellow workers and with larger machines for data analysis and data management. Working with data would be considerably easier because of graphics displays that would make the information easier to extract and understand.

Sound far-fetched? In this volume, much of what I just described is covered. The intent of this collection is to give an idea of the breadth of computer usage in chemistry and the resultant gains that can be achieved.

I thank all who have contributed to this volume, and in particular Gerst Gibbon (Pittsburgh Energy Technology Center) for his assistance in reviewing the papers.

JOSEPH G. LISCOUSKI
Digital Equipment Corporation
Marlboro, MA

May 1984

Planning an Approach to Laboratory Automation

JOSEPH G. LISCOUSKI

Digital Equipment Corporation, 1 Iron Way, P.O. Box 1002, Mail Stop: MRO 2-3/M91, Marlboro, MA 01752

Laboratory automation is, in itself not a goal, but, rather a means of achieving an objective and a process for solving some laboratory problems. That process involves a substantial planning effort. Without adequate planning, a laboratory automation project may generate more problems than it solves.

Successful projects are planned to take into account both the current needs of a laboratory and some projections as to where the lab will be two or three years out. That time period roughly matches the technology change in computing equipment and microprocessor driven instrumentation. There are two important elements that need to be included in any planning for the future: flexibility and compatibility. Will my approach provide room for growth and changing requirements (more sophisticated analysis routines for example)? Can different workstations or microprocessor systems transfer data and programs to each other?

The process of laboratory automation begins when you have clearly identified the things that you want to achieve, and why you want to achieve them. Those "things" should not be couched in phases like "I want to automate the ..." , but rather "I need faster sample turnaround", or "...more sophisticated data analysis routines will...". As far as the "why?", at some point you are going to have to justify the project, and its cost in terms of time, money, and people.

1.0 PROBLEMS FOR LABORATORY AUTOMATION

Laboratory automation can be directed at two types of problems: instrumer or experiment automation and laboratory management systems. In the first case, the computer system (microprocessor-based or larger) may be resident in the

instrument or external to it. These systems can provide us with
control of the instrument, data acquisition, data analysis, and
local storage. They should provide some means of communication
to another system, and the information transmitted should allow
you to work with the data. This implies the ability to obtain
the "raw" data - digitized spectra, chromatogram, etc. - as well
as reduced data. The technology for communications is changing
rapidly.   The practical choices today range from serial ASCII
(RS232) through the IEEE-488 bus. Eventually we can expect to
see instruments and computer system supporting the Ethernet
approach. An instrument automation system that is expected to be
around for several years, needs to be able to take advantage of
improving communications hardware and software technology.

        The second class of problem for automation or
"computerization", which may be more accurate, is that of
laboratory management and laboratory data management.   These
generally come under the heading of LIMS (Laboratory Information
Management System) systems. The concern here is usually in the
area of sample tracking, managing an archive of instrument data
and conformance to government regulations (Good Laboratory
Practices, Good Manufacturing Practices, Environmental Protection
Agency, and others). Word processing, and administrative work
(personnel, schedules, etc.) may represent added needs. In a
sense, this would be the hub of a fully automated, integrated,
laboratory system.   It should be able to communicate with the
instrument automation systems, handling the variety of data types
noted above.

        A fully-automated lab may need to contain both types of
systems.   For instrument automation systems it is important to
note that not all instruments (and experiments) can or should be
interfaced to a computer.   There are some whose accuracy or
utility can be impaired by adding an interface. With some
instruments there is also the problem of having to go inside the
device to gain access to an analog signal, that could void any
equipment warranty.   One of the choices you may have to face is
the early obsolescence of equipment due to the need for easier,
and supported, computer-to-instrument interfacing.

        Laboratory automation doesn't begin when the   first
computer is planned or delivered. Limits on your flexibility in
lab automation began to appear the day you ordered your first
piece of lab equipment.   Thinking about lab automation should
occur when you purchase instruments. How can they be interfaced?
Are there data systems for them? Are those systems compatible
with a range of computer systems or have you (knowingly or not)
locked yourself into a particular approach?

        If it isn't possible to tackle the entire job at once,
priorities can be established as to whether instrument or
management problems are implemented first.   Any system
implemented over time must have compatibility and communications
as prime factors in the planning process.

2.0   GOALS FOR INSTRUMENT AUTOMATION

What kinds of goals might we have for instrument automation?   One might be to improve sample throughput.   The level of automation involved may take the form of an autosampler to work off-hours, a robot system to take care of routine tasks, or an automatic data reduction system to capture the data, reduce it, and provide a completed report. All of these can be used to off-load lab personnel and free them for more productive assignments.   That latter point may speak to a goal of having to reduce the rate of growth of a laboratories personnel, while supporting an increasing work load.

Instrument automation may be required to provide us with more powerful techniques of data analysis and data handling; using statistical techniques that would be otherwise too time consuming to be practical; or computer graphics to gain greater flexibility in data analysis. Small data base systems of spectral libraries can help address a problem of faster component identification.

These are just a few examples where automation is a tool being used to achieve a goal.

3.0   SOME POTENTIAL GOALS FOR LIMS SYSTEMS

Here we are more concerned with data management than with data acquisition.   The goals might stem from a need to comply with government regulations and gain faster access to information.   This is a classical situation for larger (than instrument) computer systems, those capable of handling a large data base with enough flexibility to support routine and ad hoc queries, as well as exchange information with other systems. While this was noted as the hub of the laboratories automation system, it may be on a lower tier of a larger structure of, perhaps, diverse machines with differing communications requirements. For example, a well-designed and well-integrated system can help address a goal of improving information management not only in the lab, but in a plant-wide scheme for process control.

Communications through the "hub" system (serving as a data router or switch) can permit integration of data from different test stations and perform more thorough and more sophisticated analysis of the labs data.

4.0   TURNING GOALS INTO A LAB AUTOMATION PROJECT

Once a set of goals have been identified, how are they turned into a lab automation project? You can begin by setting up a measurement criteria for each goal, against which success can be tested. This point can help you in several ways. Once quantified, the feasibility of your program can be determined. It will give you a means of evaluating different approaches to the problem, and provides something to point to to show that you have met your objectives - thats helpful the next time approval is needed for a project.

Review all potential remedies. Not all situations require access to a computer to improve them. In an earlier example - sample throughput - some approaches, such as an autosampler, may solve the problem. Computer solutions, while sometimes flashy and giving the appearance of a good solution, are not always the best or most effective. They should be used only when it is clear that there is no other alternative. Why? For many people, computers and microprocessors while frequently encountered, may not be well understood. Their capabilities are sometimes overstated, and the work and knowledge required to do the job usually underestimated even by the best of people. Exhaust the simpler approaches first rather than jump into a more ambitious program. Determine if the problem you are facing is a short term (a spike in the request rate for a particular testing procedure) or a long term concern. It may be easier to live with the short term problem than rely on a project that may not move to completion fast enough.

Once goals have been stated and justified, determine a realistic time table for implementation. An urgent need may require the purchase of a "turn-key", ready to run system, rather than one in which customizing is necessary, or building one in-house. Before purchasing that turn-key system, evaluate its growth potential, and the ability to add to it without having to rely solely on the original vendor. Will it give you the expansion you need for the next few years or are you locked into what is now available?

If the solution looks to be a long-term effort, it may be worthwhile to segment it into smaller steps so that you can gain some early benefit. A project that requires a complete, instrument and management, lab automation system, might be divided into successive stages: choose some instrument automation first with lab management at a second stage, and complete the instrument work later.

This segmentation requires an evaluation of the entire problem and the individual stages in light of the following considerations:

o   compatibility - will what you do in one stage be
    compatible with the next? Will it permit easy
    integration, or will a second project be needed to
    handle it? Segmentation should make life easier, not
    create more work.

o   communications between the instrument automation system
    and management system. Does the capability exist? Does
    the software exist? Communications means more that just
    two machines having RS232 capability available. That is
    a wiring and voltage standard. Communication involves
    the useful transmission of information. In order for
    that to happen, the two systems have to agree on message
    format and protocols, error detection and correction,
    the ability to transmit ASCII and binary files, and a
    range of other factors. Who is responsible for making
    it happen? An answer of "someone" is a problem waiting
    to happen.

o   What can I expect from the vendors two or three years
    down the road? Will the equipment purchased still be
    supported then? Will they still be there?

o   Are there any changes in the vendors plans that may have
    an effect on my direction? If you are standardizing on
    one vendor, periodically review your plan with them in
    light of their development plans. Many, after the
    necessary legal paperwork has been taken care of, will
    discuss the general directions, or at least comment on
    compatibility of yours and their directions. These
    reviews may lead to adjustments in direction as a result
    of newer, and hopefully, compatible technologies.

It usually helps to have your plans reviewed by an
outside party, either a consultant or the vendors you see as
being key to your project. They may see things from a different
perspective or have encountered similar situations and help avoid
blind alleys and pitfalls. It may help to have them recommend a
solution without seeing yours. Someone fresh to a problem, and
not biased by suggested solutions, may come up with interesting
insights.

5.0 MAKE OR BUY?

Eventually, in any lab automation project, you come to
the same decision point: do I purchase a system, or have one
built to my [particular, unique, special, one of a kind] - fill
in the blank - needs? First, are your needs really that
[particular, unique, special, one of a kind]? A consultant can
help you find out. If they are, you choice is clear. If they
aren't you have some thinking to do.

6.0 CONVENTIONAL WISDOM FAVORS THE BUY DECISION

There was an interesting study published in 1975 titled "Achieving the Optimal Information System for the Laboratory" (published by J. Lloyd Johnson Associates, Northbrook Ill.). Their study is primarily concerned with hospital systems, but as far as we are going, there is no difference in the applicability of the results to general laboratory automation.

In pre-1975 dollars, the cost of a minimal laboratory system ran as high as $900,000- for a in-house developed package. That was the extreme cited, but not far behind were numbers like $700,000, $650,000, $600,000, and ranging down to $100,000. What was a minimal system? A minimal system was one "having some of the automated instruments in chemistry and hematology on-line, along with any three of the following functions":

o Test request entered through a CRT terminal or more efficient manner

o Collection list with labels printed

o Worklist generated

o Test results entered without manual re-entry of patient or specimen number

o Test result inquiry via CRT terminal

o Ward report printed

o Cumulative summaries printed

The study found that one in four will achieve a minimal satisfactory system at an average direct cost of $300,000. Another one in four will achieve a minimally satisfactory system at costs well in excess of $300,000. The development time averaged 3 years. Would a minimal system meet your goal?

In addition, some actuarial results were also reported (again from the same study):

o "Half the time the hospital will lose the total investment" o "a quarter of the time the hospital will invest more than $300,000, for a system with twice the operating cost of a leased turnkey system"

o "There is less than one chance in twenty that an in-house system will be, for a brief time, marginally better than any available turnkey system"

o "Turnkey system suppliers are constantly improving their capabilities"

7.0 WHAT HAS CHANGED SINCE 1975?

The biggest change is in price. Hardware has dropped in price dramatically. Manpower costs have risen. The power of the systems available - at a particular price - has improved, and the software is far superior to that of eight years ago. (On the established 16- and 32-bit systems, microprocessor systems, while moving rapidly, are still far behind the capability of larger vendors products.)

Hardware costs are among the more deceptive points in pricing a package, or budgeting for a project. Computers can be purchased at any price range, from $50 to $500,000 and up; with the range extending from micro to mini and maxi. The concern is not the cost of the hardware but what can you do with it? That

is governed by software. Good software can make up for the sins of poor hardware - to a point. Poor software can make a good hardware package unusable.

Don't purchase a piece of hardware and then try to find, or not finding, develop, applications software. The better approach is to find the software that will do the job, and then buy the hardware best suited for it.

Since the study was done, the range of software packages, their capability, and quality has improved. Standard libraries for data acquisition, analysis, graphics and data base management now exist. Those in the study had to write their own; frequently in assembly language rather then a high level language. Computer programming languages have evolved rapidly, with a number of them providing in one statement, facilities that took pages of code before.

With all this improvement, is "buy" still the best answer? Yes, largely because of manpower costs and the truly [particular, unique, special, one of a kind] things that need to be addressed in any organization. Duplicating an existing package means that you will have to take on the support and maintenance effort for yourself, and thats not low budget stuff! The questions noted above still have to be asked of any purchased system, - questions regarding growth, expansion, support, reliability of the vendor, and so on.

8.0 OTHER POINTS TO PONDER IN PLANNING

Where are you going to put it? Laboratory environments are not noted as being kind to sensitive electrical equipment. While the government, through the Federal Communications Commission, is working the problems of computers generating electromagnetic and radio frequency emissions that may **affect** other devices, these same devices, through dirty electrical lines, or their own emissions **affect** computers. Corrosive gases and other agents can render a machine into an expensive, though unusable, collection of metal, epoxy, and plastic. Poor electrical grounding has prematurely aged a number of people in general laboratory automation.

Many of these problems can be circumvented by carefully picking the machines location. The systems vendor should be able to give you the necessary guidelines, and if needed, make a visit to your site to look for potential problems.

Who is going to run the system? Like any other piece of equipment, computers require maintenance, updates, repairs, materials ordered,and the like. This should not be left to a committee, but rather pick someone to handle the responsibility and see that they get adequate training to handle the job.

In the course of this article, we have covered a lot of ground. The main points can be summarized easily:

o planning is essential,

o clear goals are needed with an implementation plan and
   responsibilities outlined,

o   measurement criteria for success of the project need to
    be established, and,

o   provision needs to be made for the systems growth and
    for communications.

Properly planned, a laboratory automation project can
improve a labs ability to collect, analyze, and manage data.
That planning needs to begin early in a labs lifetime, and should
include a consideration of long term and short term goals. With
that work in place, you have greatly improved the likelihood of a
successful automation project.

RECEIVED June 20, 1984

Robots and Robotics in the Laboratory: What Does It Mean?

CHARLES H. LOCHMÜLLER

Paul M. Gross Chemical Laboratory, Duke University, Durham, NC 27706

This paper addresses the limited current state- of-art in laboratory robotics and compares it to current manufacturing practice. Important questions are: "When is automation robotics?", "What is a robot anyway?" and "Where does a robot fit in a laboratory environment.?" Examples of current applications are reviewed and suggestions for future directions are presented.

The idea of a robot in the laboratory is at once a familiar and a very strange concept. Part of the problem is the association by many of the word ROBOT with a variety of ambulatory mechanical automatons of different degrees of sophistication. Currently available robots are a disappointment to many as they are neither as clever as R2D2 nor as human as C3PO of STARWARS fame. In fact, the vast majority of current robots are really arm-like machines with varying strength and dexterity; some are capable of moving hundreds of kilos and of placing such objects within fractions of a centimeter while others manipulate gram masses to sub-millimeter precision. They resemble parts of the common concept of a robot more than a whole.

Nevertheless, the previous paragraph provides the kernel of a definition for a robot. "A mechanical device which performs complex tasks with human-like skill" may be a little too general but is a good working definition. The word robot derives from the Russian for "worker" or "to work" and human work often requires significant mechanical skill. Consider then that current laboratory robots are, in essence, "blind, one-armed men" and you immediately arrive at the crude nature they possess. Current robots do not have true human skill but many common tasks are accomplished satisfactorily given their inherent handicaps.

Automation vs Robotics

How is a robot (which is used to automate a laboratory task)

different from an automated instrument (which could be designed to
perform the same task)? That is not an easy question to answer but
a reference to manufacturing robots may provide some clue.

There is a great deal of difference between a "robotocized"
production line and an automated one. Automated production works
well in situations in which the product is completely standardized
and all spatial characteristics are fixed – e.g.– bottling soda.
The advantage of robotics is in the ability to adapt to new product
characteristics – e.g.– a complete body change on the "'84 model"
in a welding or spraying operation. An "industrial" robot is a
"<u>reprogrammable</u>, mechanical device which performs complex tasks
with human–like skill". It is this reprogrammable or retrainable
aspect that makes the robot attractive from an engineering view-
point. Of course, even a robot assembly line is not completely
retrainable –i.e.– auto assembly plants cannot become textile mills
by simple software fixes. Not unreasonably, the same is true of
current laboratory robots but, especially in a routine determination function – e.g.– quality control – where the chemical "unit
operations" are very similar in procedures involving radically
different analytes, the retraining feature is an extremely
desirable advantage.

<u>Training a Robot</u>

Robots which are required to mimic complex human motion –i.e.– the
spray painting of automobiles by a 20–year veteran painter – will
require very sophisticated training utilities in the controller/-
operating system. In fact, such robots "learn" by being led
through a task "hand–in–hand" with a skilled human operator.   Such
a continuous transduction of position speed and direction into a
control program is very expensive. No laboratory robot available
today utilizes such a training scheme. In fact, current robots are
led through a sequence of steps which are individually "programmed"
by an operator to represent a unit operation – e.g.– "pour",
"tare", "weigh", "dilute", "dispense", "take aliquot" – which are
linked to become a program to make the robot carry out a particular
task – eg– "Do 100 immunoassays – Type 1". Again the difference
from automation is that the same robot can, after finishing the
immunoassays, begin a new task – "Prepare 20 vitamin assay samples
– Type 3". A real requirement for current robots is a totally
fixed coordinate system. Current robots cannot find a tube rack on
a table top, they simply go to where a tube rack "is supposed to
be".

<u>Robots: Types and Coordinate Systems</u>

Current laboratory robot operations use many of the instrument
modules familiar in conventional automation: syringe drives, relay
drivers, current and/or voltage sensors (including A/D conversion)
etc. The uniquely robotic component is a "pick and place" arm
which serves as a "mass mover" of sample, solution etc. from one
unit operation to the next. The robot controller functions to
control both the pick–and–place component and the separate unit
operations. Actually it is poor practice to separate any of the

functions of a robotic system and decide that it is <u>the</u> robotic element. It is the system that is reprogrammable or retrainable and should be thought of as an entity composed of numerous functional abilities.

Let us compare the two robot types currently available commercially for use (or adapted for use) in laboratory environments: Zymate (developed by Zymark Corp., Hopkinton, Mass.) and Microbot Alpha (manufactured by Microbot, Inc., Mountain View, CA but adapted by G. Owens and co-workers of the Procter and Gamble Advanced Instrumentation Group, Cincinnati, OH). These two robots differ in major ways each with its unique personality and capabilities. The detail of implementation has been dealt with elsewhere (<u>1</u>) and need not be dwelt on here. The Zymate is a robot specifically-built for laboratory operations and especially for sample preparation. The Microbot Alpha is an assembly robot typical of electronics manfacture but with slightly poorer positioning tolerances than the very best available for that purpose. Both are stationary robots (although the adaptation of Owens et al. translates in one dimension in some configurations) requiring precise positioning of work pieces in a circle around the workplace. Neither possess tactile or visual "sense" in standard configuration. Tactile sense can be achieved by monitoring current in the hand/finger servo systems.

The Zymate [Figure 1] moves in a cylindrical coordinate system (rotate 370°, reach 60cm, lift 56cm) under control of a microprocessor computer using DC servomotor and cable drive with potentiometric sensing of position. It possesses a "broken wrist" capable of rotation (360°) but not bending. A unique feature lies in the interchangability of the "hands". Gripper hands permit movement of tubes and other vessels while syringe hands can deliver small volumes, take aliquots and, with adapters, filter liquid samples. In some applications special hands control instument on/off functions.

The Microbot Alpha [Figure 2] is an articulated arm with a 46 in. hemispherical envelope. The arm has a positioning accuracy of 0.5 mm within the envelope. It is a stepping-motor and cable driven robot controlled by a 6502 processor that communicates via an RS-232 interface to the "outside world". Like the Zymate, it can be trained using a hand-held pendant keyboard or can be externally driven by a laboratory microcomputer. The coordinate system of a fully articulated arm is more complicated than a simple cylindrical system but this is overcome by software control. The advantage is that the Alpha can bend its "wrist" to reach into tight, angled quarters such as when tubes must be removed from a slant-tube centrifuge head.

Current Applications

Robots are best suited (in their present form) for tedious, repetitive and humanly-hazardous jobs. Tablet analysis, immunoassay determinations, polymer solubility, etc are ideal applications. Less routine perhaps, but just as tedious, are studies of enzyme action and activity which require variation in reagents and perhaps incubation timing, the "optimisation' of chemical reactions or

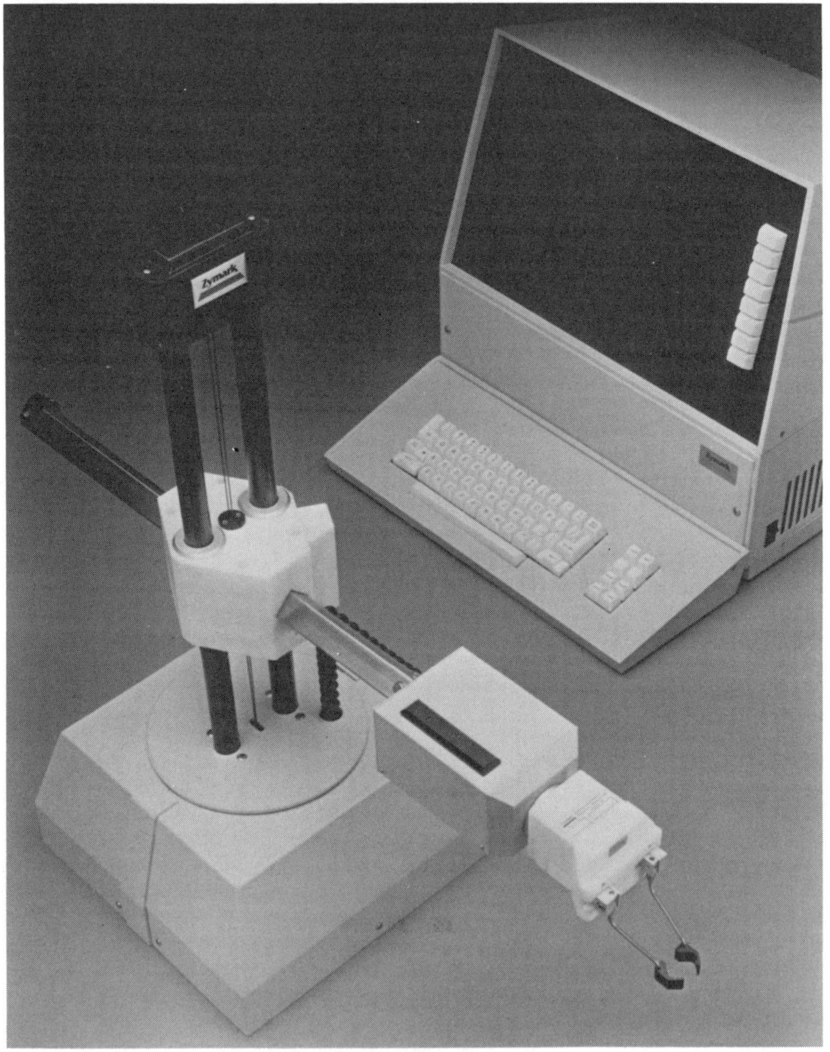

Figure 1. The Zymate (Zymark Corp., Hopkinton, MA) showing the main
robot module (center) with universal wrist and "gripper" hand attached. In the upper right is the controller with programming keyboard
and soft keys (right hand side of display screen). The soft keys
can be duplicated in a "teach/learn" pendant (not shown).

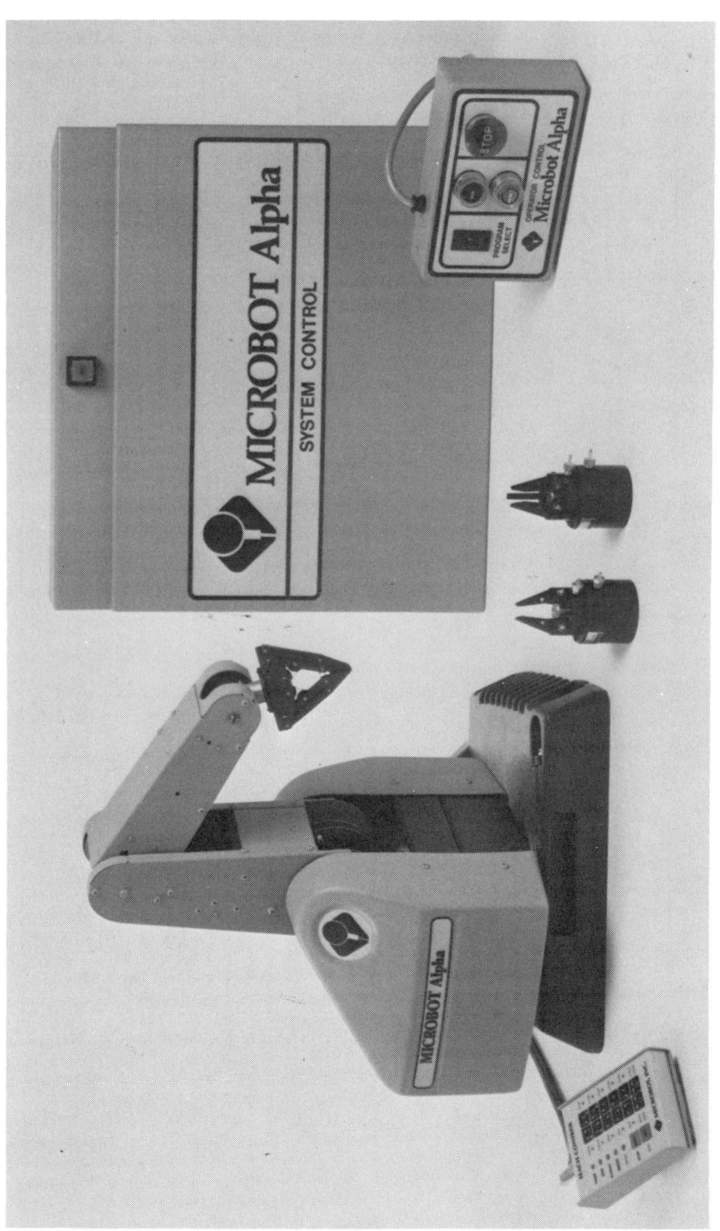

Figure 2. The Alpha (Microbot Inc., Mountainview, CA) showing: teaching pendant (lower left), robot with gripper attached (alternate grippers in front center), system controller and operator control module (lower right). Note how wrist can both rotate and bend.

routine preparation of derivatives in organic systhesis.   To date, robots do not displace people by "taking over their job" because they are used in tasks in which there is a high turn-over of personnel due to boredom.   The personnel freed by the introduction of robots can be involved in more personally satisfying tasks.   In addition, there is mounting evidence that the use of robots greatly improves both long _and_ short-time variance in the precision of quality control applications.

Immediate laboratory robot application is possible in almost any laboratory sample preparation program. Robots currently available can and do function to perform almost all of the unit operations associated with sample preparation: weighing, dissolution, centrifugation, reagent dispensing, mixing, incubation, filtering, liquid-liquid extraction and filling sample trays. All of this can be performed with complete logging of sample history.

Conclusion

There is little to conclude at present.   Robotics is an infant engineering discipline and yet the manufacturing aspect of robotic implementation is far ahead of any laboratory application. Today's robots are reprogrammable automation, they are far from being cybernauts and are hardly "clever" but their potential as an "arm" for artificial intelligence experiments cannot be overlooked.

Literature Cited

1. Analytical Chemistry A/C Interface, Vol 55, 1100A–1114A, 1232A–1242A (1983).

RECEIVED June 5, 1984

General Laboratory Data Management and Specific Laboratory Needs

W. KIPINIAK and W. FINNERTY

Computer Inquiry System Inc., 160 Hopper Avenue, Waldwick, NJ 07463

This paper describes the design and implementation of a versatile computerized laboratory automation and information management system. Discussion highlights its adaptability to a variety of laboratory environments.

Data can be entered manually or acquired automatically from spectrum generating and spot reading instruments such as chromatographs, spectrophotometers, balances, pH meters and a wide variety of intelligent instrumentation.

In accordance with good laboratory practice, information can be databased on-line or archived off-line in such a manner that allows quick and easy retrieval. Standard reports in fixed formats and ad-hoc reports in virtually any format can be easily generated either on demand or on a scheduled basis.

Because every laboratory is unique in its function and organization, this package is designed to be easily adapted to any combination of environments, products, instruments and staffing. This adaptability makes it ideally suited to any laboratory application.

A comprehensive computerized laboratory data management system has been designed and implemented successfully in a number of laboratories to provide test data collection, result reporting and total information control. The broad range of target laboratories focused the Lab Manager system design efforts on the flexibility necessary to accomodate custom configuration to the unique and evolving needs of diverse laboratories. It has been integrated into the Computer Automated Laboratory System (CALS) for complete laboratory management.

Operating in an on-line, user-friendly conversational mode, the Lab Manager system adapts to established laboratory procedures and methods resulting in minimal operational retraining and loss of

"learning curve" experience. Additionally, design criteria
included the ability to easily accommodate the strict evolving
"Good Laboratory Practice" procedures derived from regulatory
agency or in-house standards.

The system's flexibility in configuration is provided by the use
of files, called dictionaries, which contain all the tables necessary
to allow conversational on-line maintenance of user ID and security
definitions, test descriptions, product specifications, calculation
and report generation procedures. For instance, the identification
dictionary defines logon information about each user, including
name, password and security classification. The system manager
generally modifies this dictionary on-line as employees and security
considerations change. The system automatically references this
dictionary to ensure that each operator has the required authority
both to logon and execute any requested command.

System security is managed by a scheme providing for the
assignment of each command to any one or more of sixteen security
keys. Every user is assigned one or more of the defined keys and
allowed to execute only those commands which his keys enable.
Initially defined during system installation, the security tables
are maintained by assigning user identification, password and
security key. Access to the security tables is controlled by the
scheme itself - the tables being easily modified on-line, but only
by one with the required authorization.

Dictionaries

The Lab Manager system dictionaries contain all information relevant
to the unique operation of the system in each laboratory. The
fields within each dictionary record can be defined and redefined
for the specific type of data and specifications encountered within
any particular laboratory, making it possible for prompting messages
and headings displayed on terminals to vary from one laboratory
environment to another, each requesting or presenting data in
appropriate terminology. All dictionaries, required either by the
system or those dictated by site specific requirements, can be used
in retrievals and reporting as needed. The system is configured
to automatically record, for each entry in every dictionary, the
date, time and identification of the operator making the last change
to the entry.

In addition to the identification dictionary already discussed,
the test dictionary is a repository for all the tests that might
be performed in the laboratory and is configured to contain user
prompts, test protocols, quantity required for testing, assigned
testing location and analyst, standard testing time, the name of any
special calculation program to be used and cost per test. Additional
fields are added as needed to meet local requirements.

The calculation dictionary defines the procedures required to
perform calculations on test data. User friendly features include
variable declaration as either local to a test or fetched from
another test on the same or different sample, conversational input
statements and algebraic calculations. Each calculation procedure
may be referenced by any one or all entries in the test dictionary.

Each record in the product specification dictionary contains fields for the definition of tests to be conducted, test limits, product specific information and any other auxiliary information pertaining to each different type of product. Predefined tests are automatically scheduled during the sample login process. Alternately, assignment of tests and their applicable limits may be specified during or after the sample login. The number of system prompting messages issued during login of any sample may be reduced to a minimum thus providing minimum operator interaction, fewer data entry errors and greater productivity.

Complete data retrieval and report generation procedures are entered and stored as records within the database procedures dictionary. Any data retrieved from the database, or any of the dictionaries, may be formatted and printed as a report via the system report generator. This feature of the system can be used by almost anyone without any knowledge of programming because instructions are entered in English-like commands and checked by the system before being executed. Margins, spacing, headings and footings can be defined easily. Reports may be designed, printed, labelled, sorted, totalled and averaged in numerous ways. Data can be easily plotted, labelled and automatically scaled on a variety of plotters, with multi-color plotting included. The process of data retrieval, report generation and plotting is typically initiated by a single command with no further operator interaction required. All retrieval and report procedures are permanently stored in this dictionary to eliminate the need for retyping each time they are used.

Sample Tracking and System Operation

The Lab Manager system provides extensive facilities to track and control samples throughout the laboratory; indeed, management of samples and results is its primary function. The status of any sample and its associated test results can be reviewed at any time with only a moments notice. Prioritized worklists, sample status and backlogged sample reports are generated on request by any operator with the proper security definition.

Sample login involves registering a sample with the system by assigning, either manually or automatically, a unique identification called the sample ID. During this process, a "snapshot" of the product specification, test and calculation dictionaries is taken by moving all required information into the database, as defined in the configuration tables. This login process can be performed by a remote computer as well.

All samples logged into the system may require a "sampling" step before any testing can be conducted. The sampling step is optional, as defined in the configuration tables, and allows any pre-test processing, such as label printing, that may be required.

Recording and validating test data represents the single most tedious aspect of any laboratory operation. The Lab Manager system is designed to accept test results either directly from laboratory instruments or as manually entered by laboratory personnel.

Database updating and archiving functions are standard features of the Lab Manager system. Updating is performed "on-the-fly", providing retrievable data as soon as it is entered. Archiving is performed periodically to establish long term off-line storage of data from completed samples. Archived data may be recalled at any time for additional on-line analysis or reporting.

On-Line Data Acquisition and Manual Results Entry

Results of on-line, real-time instrumental analyses are posted directly to the database, with or without processing, as soon as the instrument presents the analytical data. All results are available for review and validation immediately after being posted - a feature critical to effective laboratory management. Data is acquired from laboratory instrumentation of virtually any manufacturer or function. Instruments are interfaced to the computer via analog to digital conversion, RS-232C, current loop, IEEE-4888, binary coded decimal (BCD) or bit parallel techniques.

Analog to digital conversion is performed by an interface operating at up to sixty readings per second in the +/-10 volt input range and is capable of resolving 0.3 microvolts.

Manual entry of data encompasses tests ranging from simple pass/ fail tests or sample descriptions through recording quantitative results from complex assays depending upon the configurable definition of the test. All prompts issued by the system can be quickly and easily changed on-line to suit the requirements of individual laboratories.

Additionally, the Lab Manager system will calculate and record secondary results from raw data entered. For example, a titration test would prompt the analyst for the volume and normality of titrant used and the sample weight, calculate the result, post the raw data and result to the database and compare the data to the specified limits to determine whether the test passes or fails. All results are available for review and validation, if required, immediately after posting to the database. Calculations required are structured as simple algebraic expressions and are easily specified by laboratory personnel without programming knowledge. Calculations can include addition, subtraction, multiplication, division, exponentiation and powerful intrinsic routines such as calculating arithmetic means, deviations and trigonometric functions. Complicated iterative calculations can be programmed and added to these features as required.

Testing, Validation and Sample Approval

The Lab Manager system allows setting up a sample and sequentially performing all tests associated with it. Often, laboratory procedures make it easier to set up and perform one test for a series of samples before proceeding to the next test. The system addresses this functionality through its "runsheet" processing feature which groups together all samples scheduled for the same test and displays them on the terminal allowing the operator to select samples sequentially or randomly for testing.

The validation of test results represents the practice of a
second analyst reviewing the work of another as required by various
regulatory agencies and in-house policies. The system can prevent
validation of a test result by the person who performed the test
because the identification of the analyst as well as the date and
time the test was conducted is recorded for each test. Validation
is performed on a test by test basis and includes reviewing test
results before validating or invalidating them. In either case, a
retest may be scheduled. The Lab Manager system's configurability
allows this step to be bypassed if it is not appropriate in a
specific laboratory environment.

Approving the sample involves reviewing all test results
associated with the sample and, optionally, other samples in the
database. The approval process is typically limited, by the on-line
configurable security scheme, to those laboratory personnel that are
responsible for releasing samples from the laboratory. This step
may be bypassed if it is not appropriate in a specific laboratory
environment.

Retrieval and Reporting

Reporting of laboratory data is performed by the system report
generator. Standard and ad-hoc reports are provided. The standard
reports, designed to meet regulatory requirements for documentation,
verification and control, are difficult to change whereas ad-hoc
reports can be changed easily on-line.

The most important of the standard reports is the Certificate
of Analysis. Although the report format is relatively fixed, its
unique feature is that all copies generated before or after the
official copy are labelled either preliminary or duplicate as
appropriate. This feature is essential for any laboratory that
must conform to specific standards, certify its processes and
control report documents.

Many other reports can be produced by the system on demand or a
routine schedule. The content and format of these reports are
easily established and generally defined, as required, by laboratory
personnel with no computer programming knowledge. Reports can
contain any data stored in the database. Sophisticated features
are available such as footings, headings, sorting and conditional
reporting. These report formats are stored easily and permanently
in the database procedures dictionary.

Data Networking

Laboratory instruments are interfaced to the computer by
communication loops each up to twelve thousand feet long and
supporting fifteen instruments. All instruments can acquire data
simultaneously with all laboratory management functions without
sacrificing terminal response times.

The system can be configured to run in a dual computer
environment when system up time cannot be sacrificed even for
preventive maintenance servicing. This configuration allows two

computers to access a single database at the same time providing
total system reliability. The shut down of one CPU, for any reason,
results in the second CPU assuming all essential operations of the
first, including instrument data acquisition, test data input and
report generation. The Lab Manager system can be initially
installed as a dual configuration or may be upgraded to this
capability at some time in the future. Upgrading to a dual computer
configuration is a simple and effective means to increase
responsiveness while virtually eliminating disruption to laboratory
operations by computer down time.

All data processed by the system is easily communicated to
remote computers either by the industry standard 2780 Remote Job
Entry protocol or the virtual terminal facility available. The RJE
facility is designed to communicate with remote computers at speeds
up to 9600 baud and is recommended for large data transfer loads.
The virtual terminal feature is designed for asynchronous inter
computer communication and provides the capability to access remote
databases such as CAS On-Line, Toxline and Medline. Data stored on
the local system can be sent to remote computers by either facility.

Summary

The Lab Manager system is an efficient and comprehensive laboratory
data management computer system designed and implemented to
specifically accommodate the unique operations of diverse
laboratories. The configurability and adaptability of the Lab
Manager system promotes productivity, allows creation of systems
meeting individual laboratory needs rapidly at low cost and permits
the system to inexpensively and expeditiously adapt to evolving
laboratory operations.

The cost of total system maintenance and enhancement is
currently shared by over one hundred laboratories, reducing the cost
and raising to overall system quality as compared to in-house
efforts. Proven time and time again by software life cycle, this
package is highly superior to any custom written system for long
term support and lower acquisition cost.

RECEIVED May 21, 1984

Applying Database Management in the Analytical Chemistry Laboratory

FRED BAUMANN, KENNETH A. LEWIS, and ARTHUR C. BROWN III

Varian Instrument Group, Walnut Creek, CA 94598

A general purpose, CODASYL compliant database management system is used to implement the Varian/Digital VAX Laboratory Information Management System (LIMS). The VAX LIMS runs under the VMS operating system and is compatible with the VAX family of 32-bit superminicomputers. Database utilities provided by the VAX Database Management System were extensively applied to implement many laboratory-imposed requirements. Records and set relationships were developed to meet the specific needs of the analytical environment. Ordinary programming languages are used along with the database utilities to retrieve, analyze and report data. Datatrieve, a high level database query and reporting language, is optionally available. A number of data integrity and security features are built into the system. Modification and extension of the database is possible at several levels depending on the complexity of the change and ability of the user.

Databases are used widely in commercial applications and have become the foundation of modern data processing. Various bibliographic, financial and chemical reference databases are perhaps the most familiar to scientists at this time. However, the proliferation of Laboratory Information Management Systems (LIMS) makes analytical laboratory databases accessible to most laboratory personnel. Such databases store analytical data and scientific information from which a variety of documents and reports are generated.

Analytical database design and implementation are important to the analytical chemist for several reasons:

1. The explosive growth in the amount of laboratory data;
2. The need to enhance laboratory consistency and productivity;
3. The need to share data among laboratory workers;
4. The increasing importance of data security and integrity;
5. The widening scope of laboratory automation from instruments to data management offers both opportunity and challenge to the way data is handled in a laboratory.

The interest in LIMS is directly due to the need to manage the increasing amounts of data generated by the modern analytical laboratory. LIMS systems are used in quality control and analytical services laboratories within the petroleum, petrochemical, chemical, pharmaceutical industries and others, where intelligent, automatic instruments generate large amounts of data. The laboratory must process, correlate, report and store these data securely for long periods of time.

The operating environment of an analytical laboratory involves analytical chemists and technicians generating data both automatically using instruments, as well as by manual techniques. The LIMS acquires data in several forms before transforming it finally into desired information. The LIMS may also manage data associated with products, processes, pilot plants, animal studies, toxicological studies and environmental monitoring. The laboratory manager needs records on productivity, performance, customers, accounting, personnel and inventory. This complex laboratory environment must be reflected in the database structure and consequently in the LIMS design.

The research chemist also has need for a LIMS system to store the vast amounts of analytical and other data generated in research projects. A systematic way of handling such data makes it easier to retrieve, transform and report the acquired data.

In addition to handling large amounts of data generated automatically, the LIMS database must handle data from a number of data sources: Instruments, terminals, personal work stations, and other computers. Not only does data exist in several forms but textual information such as header records, comments, reports and other documents must be accommodated. There exist well-defined relationships among the various data types in the laboratory. The dataset relationships must be carefully considered in designing the database. All data in the LIMS must be accessible by key fields such as sample number, method, instrument I.D. or laboratory. It is also necessary to support access of the stored data by ad hoc queries to extract information for correlations, summaries, retrospective studies and special reports.

Additional LIMS functions must include archiving of data, test procedures and other information necessary to meet Good Manufacturing Practices (GMP) and Good Laboratory Practices (GLP) guidelines of government agencies such as FDA and EPA. Security protection must be provided for these reasons and also to limit access to sensitive information. These requirements are stringent but not beyond the capabilities of modern database management systems.

A database can be described as a collection of inter-related data organized into records and connected by known (set) relationships. Typically, a database is organized around a function such as personnel, manufacturing, etc. A LIMS database is organized around the analytical and research laboratory. Good database design involves several well established principles:(1)

1. Data organization and storage is independent of application programs. By insulating the programs from the organization and storage of data, the users can concentrate on the "meaning" of the data instead of the physical characteristics and location of the data. Several views (subschemas) of a database are presented to the outside world depending on involvement with the database.

The non-expert user needs to only view a subset of the records, fields and sets in a full LIMS database. This is the view provided to the scientist in a turn-key LIMS system. Programmers see the database through subschemas specific to the application. At a different level, the database administrator views the complete database through the schema. Finally, the physical layout of the records is viewed by the systems programmer and the database administrator as the storage schema. These views are functional and are dependent upon the specific level of involvement with the database.

2. Data redundancy is minimized. Data redundancy is kept to a minimum by normalizing data into simple datasets which can then point to related datasets. This saves disk storage space and speeds up storage and modification operations.

3. Database schemas are centrally stored and controlled. Data definitions (schema) are stored in the centralized data dictionary. The user's view(s) of the database is defined and stored in the same data dictionary. Programs are given access to individual data fields, records, sets and areas of the database on a need-to-know basis. The database administrator creates and maintains integrity of the database schemas. The benefits of this approach are:

   A. Adjustment (tuning) of the database may be performed outside of the application programs.
   B. Programs deal with data logically rather than physically, simplifying the programming task.
   C. The database may be modified without affecting the application programs. Only those programs affected by the schema changes need to be recompiled.
   D. Database integrity is maintained in a multi-user environment through the centralized data dictionary.

4. Security protection is provided to assure data integrity. Database access is controlled to prevent unauthorized user access (for example, to sensitive areas) and to prevent unauthorized operations (for example, delete a record).

The remainder of this paper will discuss the Varian/Digital VAX LIMS and the way laboratory requirements are fulfilled using a database management system.

VAX LIMS

<u>VAX LIMS Functions</u>. A functional diagram of the VAX LIMS is shown in Figure 1. The LIMS database is organized into two portions according to function. The Data Management portion (DMDB) stores data, methods and other records related to the analytical laboratory. The Sample Management portion (SMDB) stores records pertaining to sample tracking and final results. This report deals specifically with the DMDB although the basic principles apply to both since they use the same VAX Information Architecture.

Instruments and other devices are interfaced to the VAX and the DMDB through the Data Management system. After analysis, final results are transferred to the SMDB for tracking, reporting and archiving. Final results also may be input manually from a terminal. Sample Management contains software for tracking samples and data

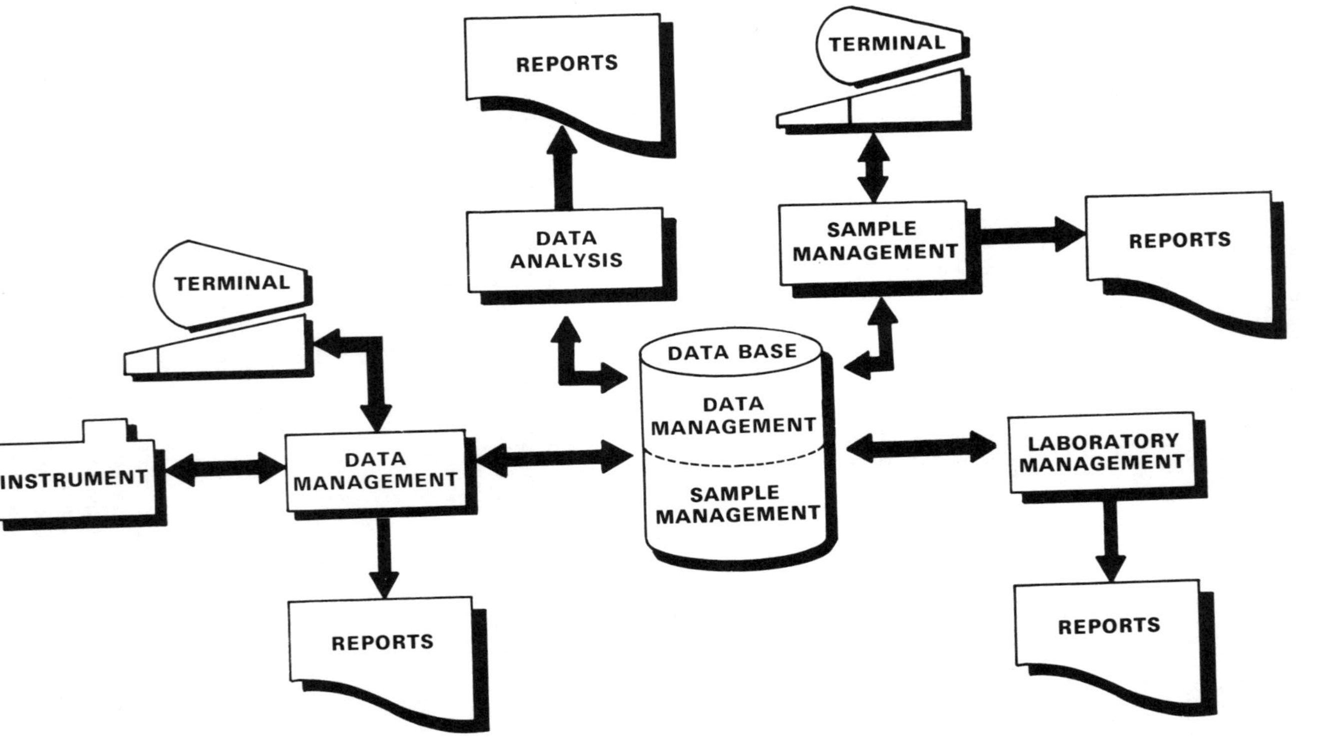

Figure 1. VAX LIMS Functional Diagram

through the processes of sample collection, login, scheduling, testing, verifying and reporting. Data Analysis modules are programs used to transform and report the data. Laboratory Management consists of a collection of software modules and reports relating to the administration of the laboratory such as performance monitoring, quality control, accounting, inventory and scheduling.

<u>VAX LIMS Architecture</u>. The above modules are application programs layered upon the VAX Information Architecture shown in Figure 2. At the lowest level is the VAX/VMS Operating System.(<u>2</u>) It supports all VAX computers in both real time multitasking and multiuser time-sharing environments. The VAX Database Management System (DBMS) is the heart of the LIMS providing the fundamental data storage and retrieval capabilities used throughout the system.(<u>3</u>) The VAX Common Data Dictionary (CDD) contains record, field and set definitions in the schema, subschema and storage schema. VAX Datatrieve is a non-procedural query and report writing language for data stored in the LIMS or other database. VAX Forms Management System (FMS) is an interactive tool to develop forms for both the entry and reporting of data, and serves both applications languages and VAX Datatrieve. Layered upon this VAX Information Architecture are the LIMS modules:

            Sample Management (LIMS/SM)
            Data Management (LIMS/DM)
            Data Analysis Library (LIMS/DA)
            Lab Management (LIMS/LM)

<u>VAX DBMS</u>. VAX DBMS is a CODASYL (Conference on Data Systems Languages) compliant, general purpose database management system based on the March, 1981 Working Document of the ANSI Data Definition Language Committee. It supplies utilities to create, maintain and use databases with complex network set relationships. VAX database utilities are summarized in Table I.

Table I.   Summary of VAX Database Utilities

UTILITY	DESCRIPTION
Data Definition Language (DDL)	Used to define the schema, security schema, subschema and storage schema
Dictionary Management Utility (DMU)	Creates, modifies, deletes or reports entities in the CDD
DBMS Operator Utility (DBO)	Used to create, modify, delete, monitor, start and stop, journal, backup, restore, recover or verify a database
Database Query (DBQ)	Interactive language used to retrieve, update and report data either directly from a terminal or called from BASIC, PASCAL, etc.
Data Manipulation Language (DML)	Data manipulation statements callable by FORTRAN or COBOL

VAX DBMS components and relationships are shown in Figure 3.
The database is composed of:

        Fields - individual data items
        Records - collection of data items
        Sets - relationship between records
        Areas - physical subdivisions of the database

A schema Data Definition Language (DDL) is provided to define the
records, sets and areas in the database. Storage Schema DDL produces
the physical description of the database records, sets and areas. A
subschema DDL produces a logical subset of the database to provide
alternative views of the database for different applications programs.
A DDL utility is provided to compile schemas and subschemas. The CDD
stores the schema, subschema, storage and security schemas. Security
schemas define the actions which users are allowed to perform on the
database. Also stored in the CDD are the Datatrieve procedures. The
Database Operator utility (DBO) allows databases to be created, modified and deleted. The CDD has a dictionary management utility (DMU)
for examining and maintaining the CDD contents.

DBMS access is provided to all VAX languages by means of Data
Manipulation Language (DML) for FORTRAN and COBOL, and Database Query
Language (DBQ) statements inbedded in the program for BASIC, PASCAL
and other VAX languages. The DML or DBQ statements are compiled
along with the application language source code. Application languages do not access the CDD following compilation. When the compiled program is subsequently executed, DBQ or DML statements request
records from or write records to the DBMS. A User Work Area (UWA) is
the buffer through which records are transferred to and from the
application programs by the Database Control System (DBCS). VAX
Datatrieve refers to the data descriptions and user procedures in the
CDD at run time. VAX Datatrieve is also callable from application
languages.

<u>VAX LIMS/DM System</u>. The LIMS/DM system interfaces instruments, data
systems and other devices to the VAX LIMS DMDB via the Instrument
Network Architecture (INA). Instruments are interfaced by storing
their communications protocols and data characteristics in records
within the LIMS database. The International Standards Organization's
seven layer open network architecture is used to separate instrument
interface problems into layers. Flexibility and simplicity are introduced since each layer deals with a simple function. The upper layers
deal with the user application program. The middle layers are concerned with routing messages between user applications and the instrument on the system. The lower layers deal with the physical routing
of messages between devices in the system. In the LIMS/DM, these
functions are performed by I/O servers and I/O device drivers. In
distributed environments, DECnet can be used for transparent communications between applications running on multiple VAX's or PDP-11's,
and can be used within INA for instrument interfacing.

<u>VAX LIMS DMDB</u>. The key to good database design is the definition of
records and the set relationships between them. The VAX DMDB schema
(Bachman diagram) is shown in Figure 4. The diagram shows the major
records (boxes) in the database and the relationship (arrows) between
the records (sets). The records and their fields are determined by
the nature of the data encountered in an analytical laboratory

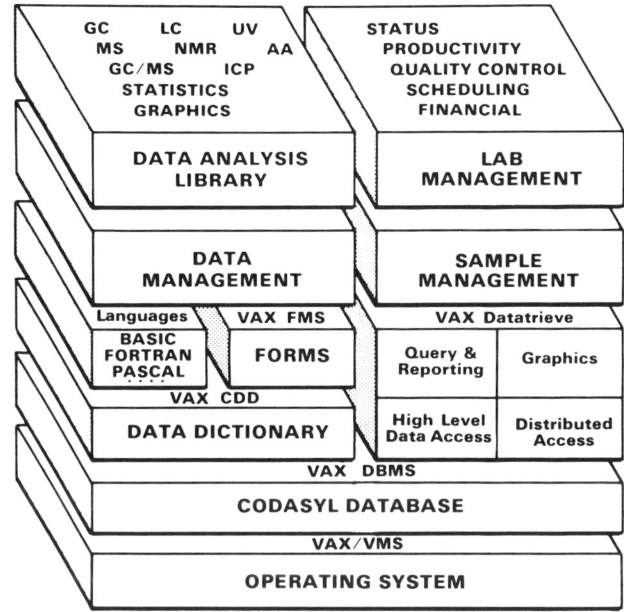

Figure 2.   VAX LIMS Architecture

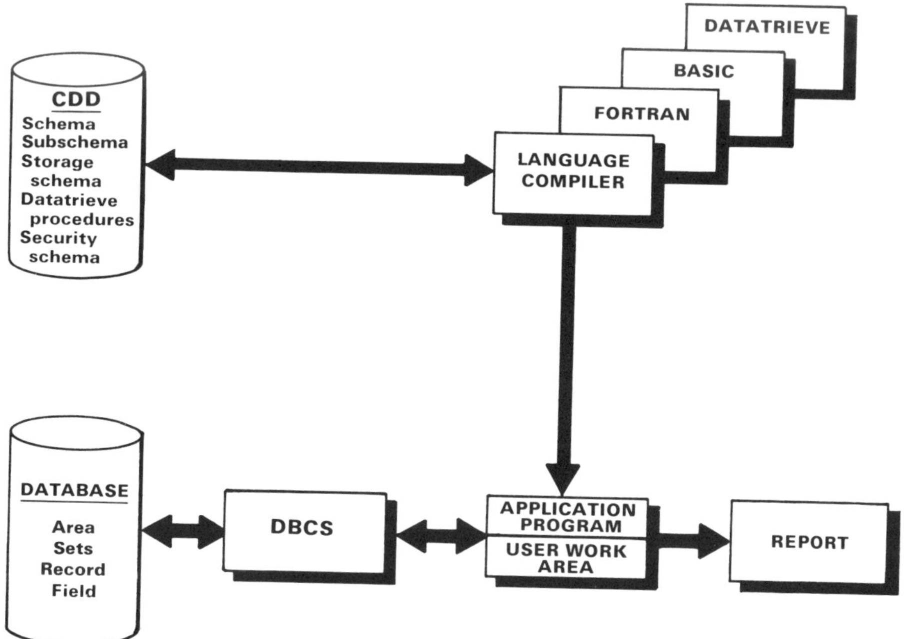

Figure 3.   VAX DBMS Components and Relationships

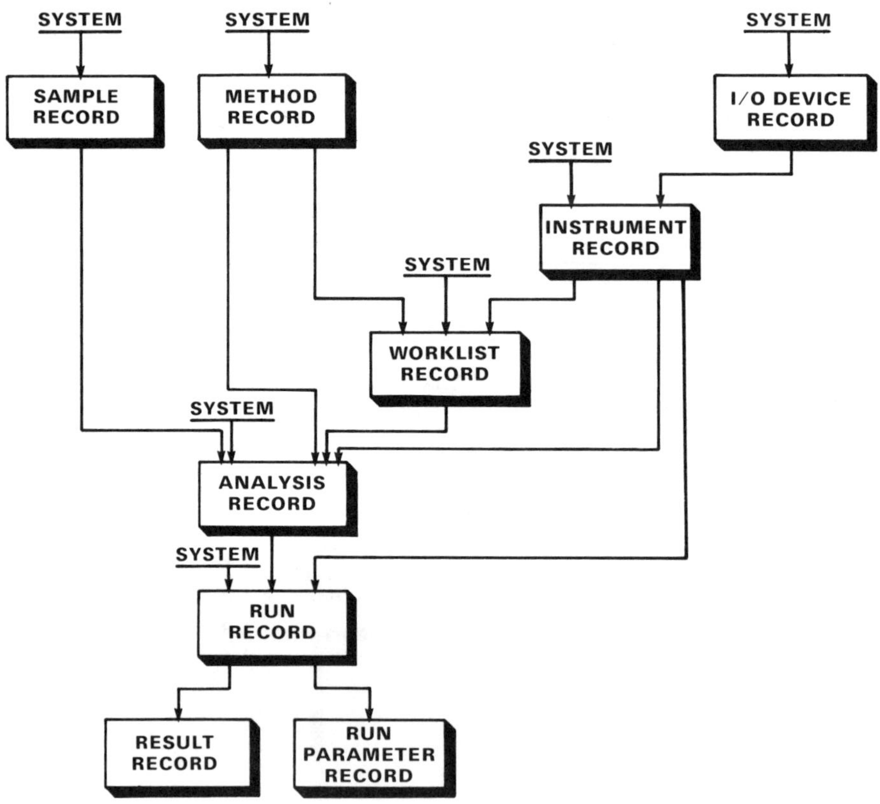

Figure 4.   VAX LIMS DMDB Schema Diagram

environment. The set relationships are determined by how accesses to the database will be handled. A direct set relationship between two records is established when a logical connection exists between them. The linkage facilitates inter-record types of queries. For example, it is easy to retrieve all analyses for a sample because there is a direct set relationship between the Sample Record and the Analysis Record. From the Analysis Record all run information for the sample can be directly retrieved and from the Run Record all results for the sample can be found. Using the database to store results for later retrieval by sample number is one of the fundamental uses of a database in the analytical chemistry laboratory.

Another important set relationship is the Instrument to Analysis Record which allows all analyses for an instrument to be easily retrieved. Since the Sample Record and Instrument Record are owners of the Analysis Record, retrieval of all analysis information for designated samples and instruments can be readily accomplished. Those records which have SYSTEM as an owner can be accessed directly without prior knowledge of its relation to other records. For example, given a worklist name, the worklist record can be accessed directly without knowing which instrument or test method it is related to.

It is extremely important to design the database with the dynamics of the laboratory environment in mind. The user must be involved to ensure the set relationships will permit the necessary questions to be asked. Poorly defined records and relationships will result in awkward programming, poor performance and, in some cases, a nonfunctional system. A few of the records are explained below.

The Method Record contains information about the analytical procedures used with instruments interfaced to LIMS. Fields include:

> Method I.D.
> Method version
> Collection procedure
> Sample storage procedure
> Sample preparation procedure
> Analysis procedure
> Calculation procedure
> Report procedure
> Sample disposition procedure
> Test components
> High, low limits for expected test results

A one-to-many relationship exists from the Method Record to the Analysis Record since one Method generally is used for the analysis of many samples.

The I/O Device Record contains information about specific characteristics of equipment interfaced to LIMS. Fields include:

> I/O device number
> I/O port I.D.
> Baud rate
> Number of data bits
> Number of start and stop bits
> Time out period

                    Parity
                    Error detection technique
These fields are used to set up the I/O drivers and I/O servers in
the LIMS/DM module.
     The Analysis Record contains descriptive information for an
analysis to be run on an instrument. Fields include:
                    Sample I.D.
                    Aliquot I.D.
                    Parent aliquot
                    In date
                    Approval date
                    Source type
                    Worklist assignment
                    Analysis priority
                    Analyst name
A one-to-many set relationship exists to the Run Record since a
sample may be analyzed several times.
     The Run Record describes conditions which occurred during the
run and comments added by the operator. Fields for the Run Record
include:
                    Run date
                    Run number
                    Instrument file name
                    Instrument file type
                    Instrument operator I.D.
Run Parameter Record includes descriptive information about the run.
Fields for a chromatographic run include:
                    Title
                    Total area
                    Remote program name
                    Drift, noise, offset
                    Autosampler rack and vial numbers
                    Injection number
                    Error messages
                    Instrument condition
                    Notes
                    Area or height flag
                    Calculation
                    Number of peaks
                    Number of unidentified peaks
                    Weight of sample
                    Weight of internal standard
The Result Record contains fields for blocks of data for a given
sample and run. For a chromatographic run, result data consist of
a series of data records for each peak:
                    Peak name
                    Peak result
                    Retention time
                    Peak offset
                    Peak height or area
                    Relative retention time
                    Separation code
                    Peak width
Raw or intermediate data such as digitized signals or area slices
from chromatographs consist of one or two dimensional arrays of

floating point numbers. The Result Record used to store data from
chromatographs and spectrophotometers can be extended to other
instruments which produce n-dimensional data by storing the points
by columns. Separating descriptive information about the run into
the Run and Run Parameter Records allows data to be stored from a
variety of instrument types.

Using The Database In The LIMS Environment

Information Retrieval and Reporting. Ad hoc retrieval and reporting
of data using Datatrieve and other VAX languages is an important
feature of LIMS as it is impossible to foresee all the future requirements for reports. The Database Query Language utility (DBQ)
is used to retrieve, update and report data from compiled BASIC,
PASCAL or other VAX languages. Data Manipulation Language (DML) is
used by FORTRAN to access the data.

VAX Datatrieve is a high level database query and reporting
language with data manipulation and graphics capability. It is a
non-procedural language intended for both the non-programmer and
programmer. A simple set of commands are used interactively and
also are callable from other languages. Guide mode can be used by
beginners to learn how to navigate the database. Remote databases
on other VAX's also can be accessed through DECnet. VAX DBMS is
dictionary-oriented and all data descriptions and Datatrieve procedures are stored in the VAX CDD.

VAX Datatrieve is ideal for ad hoc queries and low volume data
manipulations. While execution time is longer than for compiled
application languages, a trade-off needs to be made between the execution time, the cost of writing the program in a traditional,
compiled language and the frequency of running the program. A
Datatrieve report for peak data would be obtained as follows:
FOR ANALYSIS where sample ID EQ "123"
     FOR RUN WITHIN ANALYSIS_RUN
          FOR RESULT_DATA WITHIN RUN_RESULT
            PRINT RUN_RESULT
The resulting report is shown below:

PEAK NAME	PEAK RESULT	RETENTION TIME
Peak 1	123	1.0
.		
.		
.		
Peak n	456	2.0

VAX FMS provides forms management capability for application
languages and VAX Datatrieve. Forms are defined interactively at a
terminal and stored in the FMS forms library independent of data and
programs. VAX Datatrieve and FMS, used with VAX DBMS, provide the
capability to input, retrieve, modify and report data easily and
quickly.

Database Integrity. Integrity of the database must be assured particularly if the data are to be used to meet government regulations or
used as legal evidence. Several things can be done to secure the
data against user errors and hardware or software failures.
Journaling is the writing of all before and after images of modifications of the database to a journal file as well as to the database
file. The journal device should be a device other than that used to
store the database in case of failure. Database Operator utilities
(DBO) are provided to specify the after image journal device (DBO/
AFTER-JOURNAL), make backup copies of the database (DBO/BACKUP),
restore the corrupted database with the backup (DBO/RESTORE), and
reapply all changes since the last backup from the after-image
journal to the backup database (DBO/RECOVER).

Archive and Retrieve Records. The VAX LIMS/DM provides utilities
for archiving and retrieving old data. To archive, the user selects
the sample I.D.'s to be archived. The LIMS/DM ARCHIVE extracts the
selected Analysis, Run and Result Records and stores them on tape or
disk, optionally deleting them from the database. The tape or disks
can then be stored off-site or in a vault. To retrieve data from
the archive, the user invokes the LIMS/DM RETRIEVE utility which
reloads the data in the database. The user selects the sample I.D.'s
to be retrieved and writes these to the LIMS/DMDB. The user can
now access and use these records in the normal manner.

Database Security. Database security is maintained by limiting
access to the database to authorized users. Several methods are
provided by the VAX DBMS: (1) Segmenting the database into areas
and restricting access by the appropriate level of VAX/VMS file
security; (2) subschemas to limit users to those sets, records and
fields which they need; (3) a security schema which limits user's
access to the database, and also defines the transactions which they
can perform; (4) the VAX CDD restricts access to data descriptions
stored in the dictionary. Each user is granted access privileges
according to their needs. Some need only READ access to the data
for writing reports, others require READ and WRITE privileges.
Terminals also have restricted privileges. A terminal located in a
public areas may be granted READ only access for example. Off-site
dialup terminals may be restricted to use during certain hours.
Unattended terminals may be automatically logged out after a time
out period has elapsed.

Audit Trails. Audit trails are intrinsic in the design of the VAX
LIMS/DMDB. Records have creation dates, name of creator, and
comments on why the change was made. No data is over-written,
changed or deleted in place; rather, if a change is to be made to the
data, the old record is marked as having been superseded (not
deleted or modified). The new record contains all the data from
the old record along with any modifications, an indication of why
the changes were made and who made the changes. This process allows
an audit trail to be produced, sorted by sample I.D., aliquot and
test method, like any other report within the normal context of the
LIMS system. The advantage of this process is that the audit trail,
along with all the other data within the LIMS/DMDB, is maintained

and secured by the above-described access control mechanism of the
VAX DBMS. Thus, only users with proper access privileges can change
the data and then only by copying and modifying data without changing
old data (extend access). The journaling facility maintains the integrity of the audit trail as well. The auditor can roll back the
database to the last backup and see the transactions reapplied up to
any point in time.

Modification and Extensions. The analytical chemistry laboratory is
a dynamic environment. New processes, new tests, increased sample
volume, government regulations, etc., all contribute to the continual
change taking place. Improved computer systems, peripherals and
software are continually appearing and must be accommodated. The
whole concept of LIMS and laboratory automation is new and rapidly
evolving. Without the capability to extend and modify a LIMS, a
once state-of-the-art system will rapidly become obsolete.

The VAX LIMS is considered to be a basic system which can be
modified and extended to meet specific requirements. The changes
can be made at various user levels corresponding to the view of the
database.

1. Cosmetic changes to the input screens using the VAX FMS
    editor do not change the database but only the display-
    only terms appearing on the screen. Fields may be broken
    up into subfields using commas, dashes, slashes, etc., to
    make them easier to read. Fields can also be surrounded by
    boxes, set in reverse video (black on while) underlined,
    colored, made to blink, made double height, or displayed in
    bold. Such changes make the forms easier to use and friend-
    lier.

2. Comment fields, generic test result records and parameter
    records are included in the database. These fields and
    records can be used without recompiling the database to
    store data, instrument parameters and comments which were
    not explicitly defined in the original database. The actual
    format and use of the data and parameters is determined by
    the application programs which use them.

3. The database subschema entities such as set names, record
    names, or field names can be renamed without changing the
    data or the set relationships by means of the ALIAS feature
    in VAX DBMS. This is a useful feature for renaming fields,
    records and sets to suit a particular laboratory environment.
    For example, the term 'sample' or 'specimen' may be preferred,
    depending upon whether the laboratory is in an industrial or
    hospital environment. ALIAS is also used to create LIMS
    subschemas in foreign languages. This is done by making a
    copy of the subschema using the DBO/EXTRACT utility, adding
    the ALIAS entry, and compiling the new subschema using the
    DDL/COMPILE utility and DBO/MODIFY utility. Only those pro-
    grams using the ALIAS need to be modified and recompiled.
    The database does not need to be rebuilt.

4. Although more complicated, the schema may be edited to add
    new records, or fields, or create a new set relationships
    within the LIMS database. This function is usually done by
    the database administrator, a systems programmer or system
    manager in the role of database administrator. VAX DBMS

provides a Database Administration Manual for this purpose. (4) The VAX database utilities summarized in Table I
provide facilities to develop schemas, create CDD direc-
tories, and create, modify and use the database.
After the schema and storage schema have been modified using the
text editor to add the new fields, records, or sets, they are compiled
using the DDL/COMPILE utility. Next, the database is modified using
the DBO/MODIFY utility. Old subschemas are still operative and old
application programs which do not use the new fields, records or sets
may still be used. New subschemas are created to use the new entities and these are used by new application programs. One cannot delete or modify anything from the old schemas or subschemas or add
new areas. This allows old programs using old subschemas to continue
to run without recompiling and rebuilding the database.

The LIMS database is amenable to changes in application modules
such as instrument interfacing, data analysis and laboratory management. The independence of the data from the programs is a major advantage of database systems. A carefully designed system with built-
in tools and utilities to provide easy modification and extension is
the best solution for a system configurable to the user's environment
and capable of fulfilling future needs.

<u>Literature Cited</u>

1.   Martin, James In "Computer Database Organization"; Prentice-Hall,
     Inc.: Englewood Cliffs, New Jersey, 1975.
2.   "VAX Technical Summary", Digital Equipment Corporation, 1982.
3.   "VAX-11 DBMS", Digital Equipment Corporation, August, 1982;
     Vol. 1-3.
4.   Ibid., "Database Administration Manual", Vol. 1 ADJ966A-TI.

The following are trademarks of Digital Equipment Corporation:
Datatrieve, DECnet, FMS, VAX, VMS.

INA is a trademark of Varian Associates, Inc.

RECEIVED June 5, 1984

Network and Communications

DOUGLAS ST. CLAIR

Digital Equipment Corporation, Stow, MA 01775

Frequently the data communications solution is addressed independently of other organizational communications needs. The knowledgeable user considers Data Processing Networks a PART of the communications problem and rightly so. The entire communications for the organization consists of voice, FAX, written, etc. A large number of new services are becoming available on the "computer network" that augment or replace existing systems. Interoffice mail for example has an analogue in the Electronic Mail systems available on computers. This paper will address various aspects of computer networks independently of the total communications solution for an organization. But, thought should be given to integration of all communications in your organization based on available resources and your unique needs.

There are two primary definitions for the term Network. One definition for a network describes a few computers with lots of terminals. Such installations frequently have grown from a batch environment with a large central mainframe computer to which terminals have been added. The term data communications is used to describe terminal (or terminal like) communications between the terminal and host. This is perhaps the most mature communications area. It was developed initially to support Teletype Equipment over telephone lines. Once computers began to support terminals this technology was a natural adaptation.

The second more modern definition describes a Network comprised of more computers tied together with relatively few terminals on each computer. This second form of network became a reality with the mini computer revolution was spurred by acceptance Digitals PDP-8 and PDP-11 Mini computers in the 1960s and development of Digitals Network Architecture (DNA) in the 1970s. In order to clarify the distinction lets call the first form of "network" Data Communications and the second type a Digital Computer Network.

A primary consideration for networking is resource sharing. The newtwork is clearly becoming more and more like what was formerly the system. The various functions once located with in a single system (mass storage, computation, printing, etc) are now being shared and called file servers, compute servers, print servers, etc. If resources are cheap then they can be replicated. If they are expensive resources must be shared. This law of economics dictates putting computing power at the desk and sharing printing, and storage. The need for storage is not related to computing power. Laboratory problems with trivial computational need can acquire large amounts of data. Storage costs are not as amenable to price reductions as CPU power. Therefore the network should offer the economy of large disks as a shared resource. An efficient networking scheme would also allow the user to either move the entire file over the network or only the records on interest from the file. The second approach, moving records, is apt to be preferred because it minimizes total storage requirements (only one copy of the file in the net) and simplifies procedures since the one copy is updated and there is no question about getting and "old" copy.

Level 1

Central. This highest level provides resources that are viable in terms of their cost when the benefits are realistic when spread across the entire organization. It is also the level at which information which is of greatest sensitivity should reside.

Level 2

Division. This level merges the resources necessary to support two or more departments. The emphasis is moving toward more managerial functions. However, it is also the logical level for resources whose benefits are such that the cost is only realistic when spread across all lower levels.

Level 3

Laboratory. The level provides support to more than one group. It is assumed that data collected by several groups within the department is necessarily related and therefore resources at this level support collating and analyzing data from several groups. In addition the is the lowest level at which administrative, financial, and management functions reside.

Level 4

Department.   A department is large enough to contain
more than one group.   Something on the order of seven
members as a practical minimum.   The department is the
level at which administration and management become
required.

Level 5

Group. The group is the smallest unit that can function
with essentially no administrative overhead.   Groups
therefore consist of approximately 3 members.
People at this level are implementors. They perform the
tasks associated with collecting raw data and laboratory
analysis. The only function to probably bridge between
group and department would be clerical.

Project or Task.   This is the smallest functional unit
in the organization.   One person could be responsible
for several projects.
        There are three primary measures of performance for
a communications channel.   They are distance, speed, and
cost.   Speed and distance are conversely related to one
another.   That is you can have speed or distance but not
both Cost is directly related to both.   That is to say
if you want either speed or distance you are going to
pay for it.
        Unfortunately the relationship is not linear and
increases in speed result in dramatic increases in cost.
The   cost   distance   relationship   is   not   quite   so
dramatically non-linear.

Laboratory Front Ends

The laboratory front end device looks conspicuously like
a personal computer to a network.   In fact there are
other interesting parallels between the growth of
terminals to personal computers and the evolution of
laboratory front ends.   At this time the most common
interfaces are RS-232, RS-422, IEEE-488, and parallel
interfaces.   Many of these are clearly adoptions of
terminal interfaces made to adapt these devices to
"automatic data entry".   However, these interfaces were
never designed to allow for Networking.   They were
intended to make the devices look like dumb terminals to
the host.   But the advent of micro processors and their
implementation in laboratory equipment clearly indicates
their direction continues to parallel the path followed
by the terminal toward a personal computer.   They will
very quickly require the same services that personal
computers require.   The next generation of Lab devices
will have increased mass storage to fuller utilize the
intelligence buried inside.

Personal Computers (Disk Based Systems)

A significant difference in communications requirements is expected when storage media is distributed along with personal computer systems. These configurations introduce a requirement for networking (host to host) as opposed to data communications (terminal to host).

File Transfer And Task To Task Communications

Personal computers will require file transfer and task to task communications traffic from the network. The characteristics of networking are therefore marked by considerable increases in both the speed and quantity of data to transfer. In addition it is not the speed and power of the computers but the size of the storage at the ends of the network link that dictate networking requirements.

Mass Storage Devices In Laboratory Systems

The inclusion of mass Storage Devices in Laboratory instruments. The laboratory front end manufactures are considering introduction of mass storage into their devices. The backup of these devices is generally assumed to take place from Winchester Disks to floppy diskettes o streaming tapes. A cop out solution is to add a second Winchester and produce a shadow volume and hope both disks don't go belly up at the same time. However, a communications link to a larger host is clearly a vastly superior solution to any of the local mass storage approaches (floppies, streaming tapes, shadow volumes).

Problems Associated With Local Storage

In the case of floppy diskettes the backup procedure is clearly cumbersome. Archival storage on streaming tape is an unknown quantity. It is necessary to handle 1/2 inch tape every 18 to 24 months to avoid print through and mechanical problems. It is not clear that streaming tapes are not immune to the same problems and will probably require the same type of handling. This handling is at this time not supported either by equipment, knowledge, or procedures for the streaming tape media. Local backup of Winchester disks to mass media also offers the prospect of incredible labor costs with the proposed proliferation of these devices.

A Networked Solution To Distributed Storage

If the appropriate communications link existed then there is tremendous potential for its application in a large number of these Winchester based systems.

Networking Effects on Data Manipulation

A Digital Computer Network allows the users to place an impressive amount of computer power anywhere. In the research environment this allows the researcher to control the experiment, verify the data, extract and record at the experiment site. The dramatic reduction in the cost of data manipulation allows one to put many computers where ever they are needed.

Networking Effects On Data Access

The researcher can store the data at the point of capture or send the information over a Network to a larger department level machine. Since the cost of storage is considerable relative to the cost of computation every attempt should be made to minimize this cost. The cost of networking tends to take advantage of the same technology as data manipulation therefore the transport of information over the Digital Computer Network to larger more economical storage on department level machines makes sense. In addition the use of a small machine on site allow the amount of data to be reduced by selecting the data before transmission. Also use of the departments machine to centralize the costs of backup and restoration of data prevents errors caused by a scientist trying to do a computer operators job.

Wide Area Networks Characteristics (Distances Greater Than 3000 Meters)

Wide area networks are defined as those utilizing the services provided when the user does not control the channel. For example few users can purchase the rights to install a wire from a facility in Boston, MA to connect a facility in St Louis, MO. The same technology is generally used if you need to communicate across town or across the street you buy it as a service. The divestature of ATT may produce a step function in long lines costs in the near future. The costs of wide area network services from the telephone company have caused a number of large users to stop using these services and buy satellite communications to replace long lines and broadband networks to replace the service with in a campus like facility. Cost of these services is regulated and set by tariff. Recently the operating company in Oklahoma requested a rate increase for residential telephone service attaching personal computers to public data base services. The cost of connection would increase the basic rate approximately 5 times the voice only rate there is no apparent change in the actual service being provided. A similar rate filing is being made in Texas.

Local Area Networks

These networks are limited to distances greater than 45
meters and less than 3000 meters. Examples of the
techniques employed include, PABX, Ethernet, Broadband,
Fiber optics, and Microwave.

Clusters

This distance, less than 45 meters, is roughly 3 times
the distance minicomputer internal busses ran in the
1960s. A bus is the name for the communications channel
inside the machine. Data rate for the Ethernet is 10
million bits per second.

Broadband

Connecting terminals to hosts appears to be the most
widely used area for broadband. Allows the same channel
to carry voice and video signals. The signal rates for
broadband are not impressive for data communications at
this time under 1 million bits per second for most
applications.

Fiber Optics

This technology is attractive from a number of
standpoints. Everyone would like to see coordination of
the material and connectors so that migration in the
field could take place as follows. For example a
customer obtains a fiber optic terminal to host link.
Then replaces the Terminal with a Professional/Personal
Computer. The single Professional/Personal Computer is
replace by a cluster of several Professional/Personal
Computers. The cluster expands to contain a machine of
the VAX class which is connected to the
Professional/Personal Computers by Ethernet and into a
high performance cluster (a very high performance
Cluster of large machines) via the same Fiber Optic
Link. At each stage the terminal through High
performance cluster the same fiber optic cable could
handle the traffic. Planning ahead the right fiber
optics material will allow migration such as this to
take place.

Historical Development Of Laboratory Communications

The development of Laboratory Devices has proceeded as
follows.

Phase 1 Manual Data Entry. Data is keypunched manually
and entered via cards or paper tape to the computer.
Later Data is entered via a terminal to the computer.

<u>Phase 2 Analog To Digital In The Host</u>. The laboratory manufacturers follow by connecting their devices directly to the data processing system. These Laboratory devices provide and analog signal. The Laboratory Device relies entirely on the host for intelligence. The host provides both control for an analog to digital (A/D) Converter and processing of the data. A moderate improvement in efficiency is achieved by multiplexing several Analog signals to a single A/D in the computer.

<u>Phase 3 Digital To Host</u>. A/D conversion moves into the laboratory device and BCD or ASCII digits are sent in serial or parallel to the host. The laboratory device now looks like a terminal. The controller operates in character interrupt mode. Logical extensions of this technology have followed the manner in which we interface terminals. The first interface is a single card interfaces and then multiple interfaces per card. However, both are interrupt driven and considerably load the CPU. The next logical step is a DMA or silo controller. There are "standards" the RS-232-C (in many variations) and the IEEE-488.

<u>Phase 4 Intelligent Lab Devices</u>. Laboratory device manufactures offer intelligent devices with dedicated CPUs. To do this they have developed software expertise These laboratory devices relay on CPU manufacturers hardware and software. The laboratory device manufactures include a micro processor chip in the device with sufficient power to do the job that once took a PDP-8. In addition the machine trouble-shoots the process and offers remedial action to the operator. This substantially changes the communications interface. The data moved to the host is substantially pre-processed. The analyzer is now capable of sending "packets" of pre-processed data to the host for more general analysis. Increased storage is also made available. The analytical device now begins to look more like a disk than a terminal in terms of potential data transfer rates.

<u>Phase 5 Multiple CPUs And Intelligent Laboratory Devices On The Network</u>. Computer manufacturers have developed systems with multiple CPUs and multiple operating systems (i.e. A Network). The construction of networks comprised completely of intelligent laboratory devices is a real possibility with the availability of the Ethernet Specification. Small local area networks of laboratory deices will develop using Ethernet and other technologies. Later these local networks will be interconnected probably via routers and gateways to broadband, Satellite, and telephone facilities.

<u>Conclusions</u>

The path the data processing manufacturers have taken is
being followed by the Laboratory device manufacturers.
The laboratory devices are becoming increasingly
intelligent. This trend implies that the personal
computer may have come too late for many instruments.
They will use the chips instead. Manufacturers are now
integrating chips into their devices and will soon not
be satisfied with less than true host to host
communications capability.

RECEIVED July 13, 1984

Introduction to Graphics

JOSEPH G. LISCOUSKI

Digital Equipment Corporation, 1 Iron Way, P.O. Box 1002, Mail Stop: MRO 2-3/M91, Marlboro, MA 01752

```
Graphics -- (noun) the science or art of drawing,
particularly of mechanical drawing, or of drawing
to mathematical rules.  (Britannica World Language
Dictionary)


Computer -- A device capable of accepting information,
applying prescribed processes to the information,
and supplying the results of these processes.  It
usually consists of input and output devices, storage,
arithmetic, and logical units, and a control unit.
(Computer Dictionary, Sippl and Sippl)
```

Taken together, those two words describe both a body of knowledge and a tool for applying the "rules" to information and manipulating and it. In a broader context, computer graphics is a endeavor that not only deals with rules and data, but also encompasses the means of displaying that information and interacting with it.

1.0 GRAPHICS APPLICATIONS

The applications that the field embraces can be grouped into four categories:

0097–6156/84/0265–0045$10.50/0

o   Data Representation - data plotting packages that
    permit us to visualize the relationships between
    variables,

o   Modeling and Line Drawing - the representation and
    display of real or imagined objects,

o   Image Processing - The display and analysis of
    real or imagined objects. This would also include
    the enhancement of information about those
    objects.

o   Document Preparation - combining text processing
    and graphics.

Over the next few pages, we will present a description
of these classifications and the particular hardware
and software for their use.

1.1   DATA REPRESENTATION

1.1.1   Graphs And Charts -

Data plotting is one of the most common laboratory and
commercial applications of graphics. Most of us
became acquainted with the topic in high school
algebra and science classes. The basic problem is to
pictorially present (as a species we can grasp
pictures much more easily than lists of numbers) the
relationship of two or more variables. This usually
involved deciding the best way to view the data
(barcharts, piecharts, or line graphs, for example)
scaling the figure, adding the data, labeling, an so
on.

This is a straightforward problem, for us, with a
pencil and paper, but no so trival for a machine whose
strength lies in manipulating discrete integers.

Lets take a step back and look at the process. Say
that we are plotting a simple scatter plot of some X,Y
pairs. The first step is to pick a suitable graph
paper, there are a lot of them, different types and
scales - someone else has gone through the trouble of
working out the rulings, line thickness and such and

printed the pattern on a writable (and fortunately
erasable) surface. They left some white space on the
edges for labels and notes. Thats not the case on a
CRT (or pen plotter). The best we can usually hope
for is that the display surface will have a
rectangular coordinate system. The lower left corner
will have a particular set of coordinates and so will
the upper right corner.

Given this situation, the plotting program must
allocate the actual plotting region - allowing space
for labels - and then set up the grid according to the
users needs. That process is not difficult if we are
dealing with linear axis (drawing a box, figuring out
where the tick marks should go, what line patterns are
needed for major and minor ticks) but can become very
complex when non-linear axis are involved. It now has
to take into account scaling functions (to account for
the non-linearity) as well as scaling factor (to make
sure that the winds up in the graph paper rather than
in the margins.

Next, we deal with the scaling of the paper. Again
for us, it is a minor matter of working out a
convenient scale. We have so many labeling positions,
the data covers a known range and we can pick some set
of labels that are easy to work with - increments of 5
or 10 might work out well for a particular case.
Computers don't understand the word "convenient".
Given a set of numbers and told to scale them you
might wind up with values (at the minimum and maximum)
of .237 and 9.314 with increments of .908 if we have
10 major tick marks. Not very "convenient" if you are
trying to read information from the plot. The program
needs to have sufficient intelligence to choose an
easy to work with set of labels - a complex problem
since we first need to agree on the definition of
convenient and then add the constraint that it not
waste a lot of the viewing surface by picking too
broad a range.

Adding the data to the scaled plot is the least
troublesome aspect of the task and can be handled
easily by most plotting programs as well as people.

The problem becomes more complex as we begin working
with bar-graphs, piecharts and more involved plotting
applications.

The point of this excerise is to illustrate that
simple graph plotting problems that would not tax a

high school student, are far from simple for a
programmer to anticipate, while he designing a package
for general purpose use, when you consider the steps
that are involved - and taken for granted. This is
the basic reason why computer graphics systems -
particularly software - are expensive and not easily
produced.

The use of data plotting can vary widely. From a
researcher trying to display an acquired analog signal
to response surface plot that might be used a
publication or presentation. The figures that follow
(figures 1, 2, and 3) are some simple applications
produced with one scientific graphics package (Digital
Equipment Corporations VAX-11 RGL).

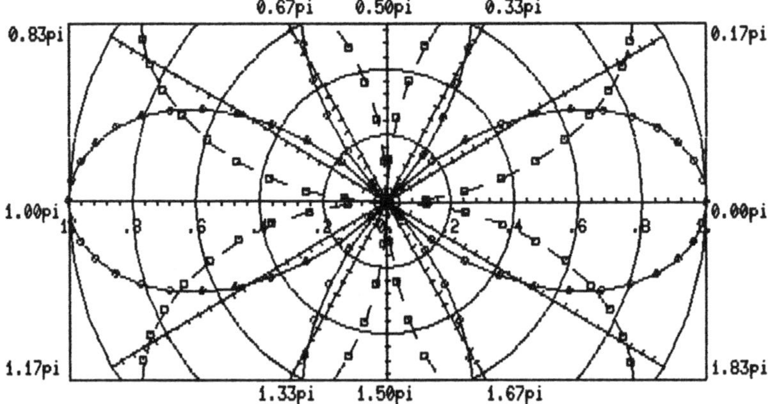

Figure 1. Polar coordinate grid. Papers of this type can be used
to show the radiation pattern of an antenna, or the EMI/RFI emissions
for a VAX computer or terminal under test for FCC compliance.

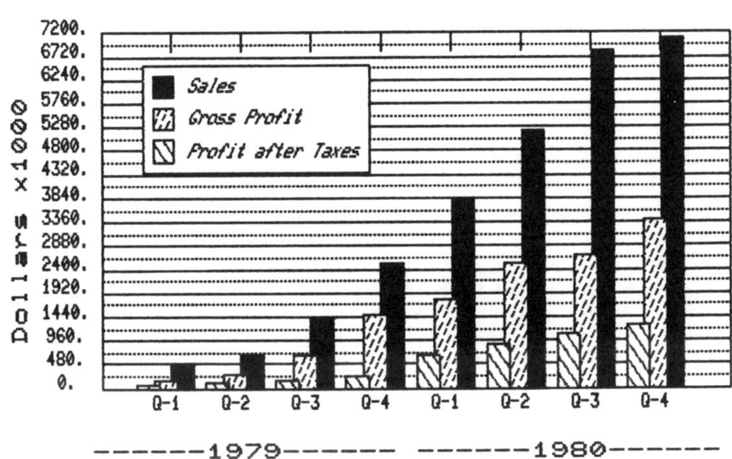

Figure 2. Different sets of financial data can be more easily understood and compared graphically.

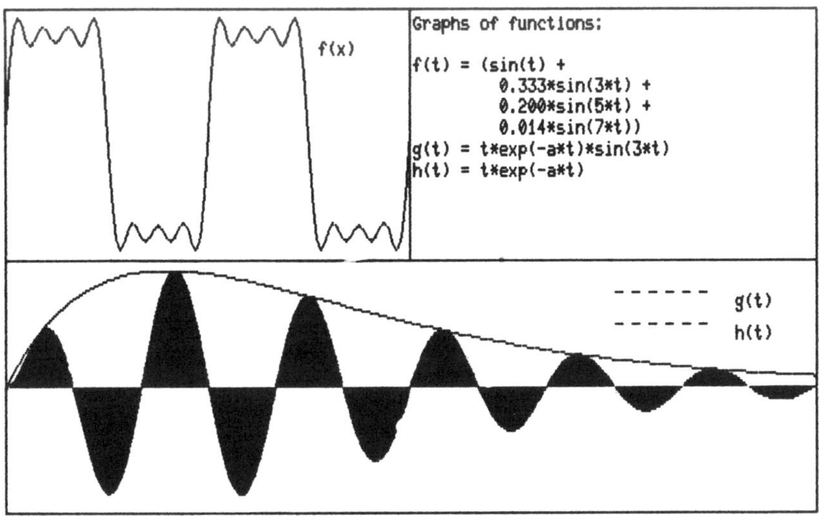

Figure 3. Different viewpoints can be used to show data or descriptive text.

1.1.2   Hardware Requirements -

The display requirements depends on the users need and
thus can vary over the spectrum of resolution, number
of colors, and display type. Interactive users will
need a CRT - the only form of an erasable graphics
medium. The resolution of the device - and its
sophistication - simple plots (one or two per screen)
can be done on a low to medium resolution device (on
the order of a - 768 x 240 addressable points). As
the graphs become more complex or more numerous on a
single screen, the need for higher resolution (1024 x
1024) increases. Not only because of the need to put
up a lot of distinguishable points, but for text used
to label and annotate the graphs. Many terminals use
an   8 x 10 character cell for its normal size
characters. The half size characters (available on
some graphics terminals), on those same devices are
barely readable. If a lot of text and graphics is to
appear on the screen high resolution is need simply to
have the necessary number of dots available to write
small characters that can be understood.

If the user wants to walk away with his graph, he
needs some form of hardcopy unit. The nature (raster
hardcopy, raster printer, or pen plotter) depends on
his use for the copy. A screen copier might be used
to produce suitable overheads for an informal
presentation; and a pen plotters' output might be
better for a speech given to a professional society or
funding agency (they may be judged by the quality of
his presentation as well as its content).

1.1.3   Other Forms Of Data Representation -

While graphics and charts are the more common forms,
there are other approaches to illustrating data - many
of them are only practical when generated by a
computer. Lets take the pie-chart as a starting
point.

The segmented circle might be an appropriate format
for displaying the proportion of our business that is
generated by various market segments, its utility is
limited to only a few variables. Use too many
divisions and labeling becomes a problem. You might
wind up obscuring more information rather than making
it clearer. The oil companies solved that problem
neatly when they wanted to show the proportions of our
total oil imports from various countries in one

display. The number of data points is large, so
rather than use a pie-chart, they took a map of the
world and then, dynamically, distorted the size of
each country so that it was proportional to the amount
of oil we imported. An example of such a display on
another topic can be found on the following page
(figure 4, taken from the Boston Globe for March 7th,
1982).

The concept of the bar-chart has also be extended
through the use of computer graphics, extensions that
are particularly effective when the item being
measured is a function of political geography - states
and towns. The illustration on the following page
(figure 5) comes from Harvard University. The
sequence of nine frames shows the population of the
United States as a function of time (the height of the
contour at any point is proportional to the population
in that area).

The Laboratory for Computer Graphics and Spatial
Analysis (Harvard Graduate School of Design) has a
system called ODYSSEY. A combination of data base and
graphics display, the ODYSSEY system can represent
social and economic statistics as a function of
political subdivisions. For example the amount of
agricultural land in Massachusetts (by town) could be
shown as a map of the state, with the outline of a
town projected above the background by an amount
proportional to is agricultural land mass. Color can
also be used to perform the same classification,
although the segmentation would not be a fine due to
limitation on the number of colors available in the
printing process. This same set of techniques has
been used to show voting patterns.

An "Animation Information Retrieval" package by the
same group can be used to shown the relationship
between multiple variables. One example they cite is
the pattern of airline traffic, arrivals and
departures from U.S. airports, both as a function of
time of day and geography. The animation gives the
viewer an easy grasp of the data where more
traditional forms of display (including lists of
numbers) might leave you a bit bewildered.

The appreciation of multi-variate relationships need
not be as involved as it is in the Harvard package.
There are some relatively simple techniques for
illustrating data. Take a familiar figure, a face or
truck for example. Let a copy of a figure represent
an individual product line in Digital. The sizes of
the noses might be proportional to the amount of
revenue obtained from government sources, the ears the
university segment and the size of the eyes the
industrial income.

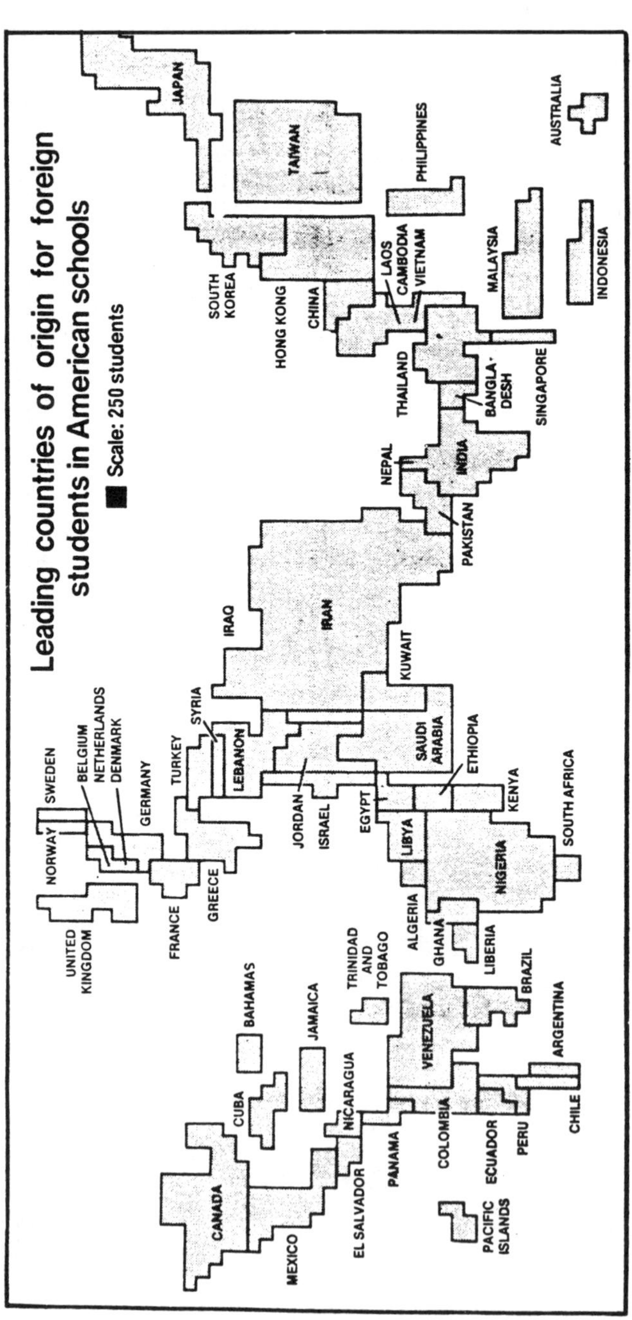

Figure 4.   Example of a modified pie-chart display.

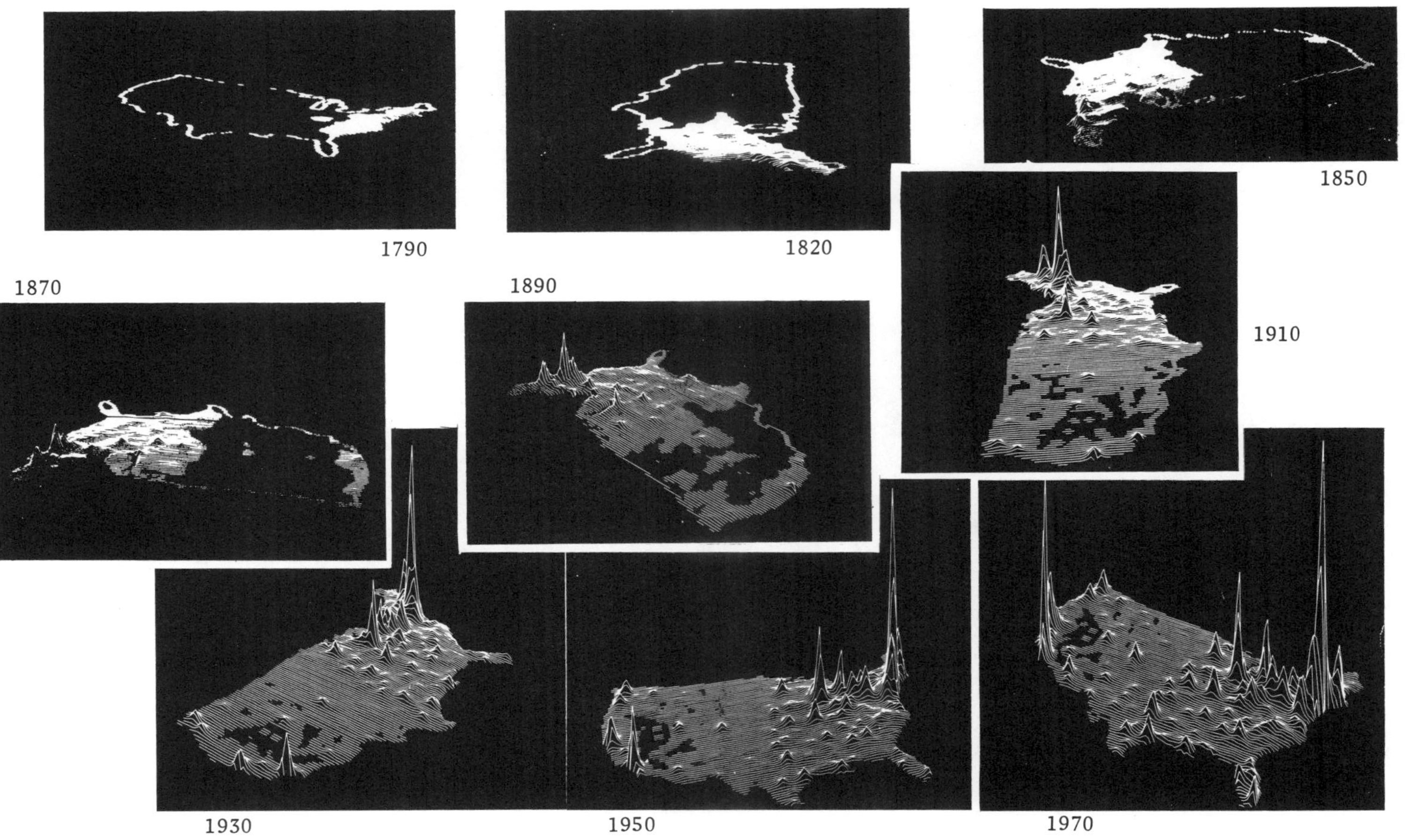

Figure 5.   Population growth in the United States from 1790 - 1970.

1.1.4 Hardware Requirements –

The displays used in the last section are usually raster devices due the the need for a variety of colors or shading options. The resolution of equipment is high so that the figures and shapes on the screen can be accurately represented without distortion. Some form of hardcopy is almost a given. The type will depend on the nature of the presentation. 35 mm slides might be appropriate (a Matrix system [camera attachment for screen copying] for this purpose could run about $10,000) or computer output on film could be used for animation work. Plotters are useful where the emphasis is on line drawing, with little filling of areas due to the amount of time needed to complete the operation.

1.2 Modeling And Line Drawing

Modeling and Line Drawing applications pretty much cover the map of end users. One common reference is CAD – Computer Aided Design. The uses here range from circuit board layout to automotive and aircraft design. Stone and Webster had a package that was running on the PDP-15s for architectural work. That program allowed you to draw a building or room – in three dimensions – and walk through it. One demonstration showed an auditorium. With the package you could "stand" in the back and view the stage, or "stand" on the stage and see how the audience would be arranged.

The visualization of molecular structures finds a home in this realm. Using packages such as TRIBBLE (DuPont) or the PHOPHET system (Bolt Baranek & Newman, Cambridge, Mass.) one can describe a molecular structure and have the system display the three-dimensional structure as it would appear when viewed from various reference points.

Artificial Intelligence, urban planning and the automotive industry use this form of graphics to represent and view objects. The design of a building complex might be represented in the machine an shown on a color monitor so that the designer could "walk" around the structure and view them from different perspectives.

Advertising is a common use. The CBS and ABC television logos are the result of computer generated graphics. The coloring, shading and dynamics are machine generated - though not in real-time. The display media is film and image is drawn with light, one frame at a time. The same approach was used in the past year to advertise an FM radio station. During the time slot, the viewer was taken on a simulation of night-time car ride. And speaking of simulation, flight simulation is an important one - the landing of a jet on an aircraft carrier for example.

Regardless of the application, the elements of such a system can be broken down as follows:

> Data entry devices - this is the means of describing the object of interest. It can be a mathematical function, or a digitizer that provides a set of coordinates (two or three dimensions depending on the problem) that describe the object.

> A data base used to store and retrieve the objects description.

> A software package that can be used to generate the display. Depending on the needs of the user, the package might have hidden line removal [the ability to not display lines or surfaces that would not be seen if the model were a solid body] plus the ability to rotate the object about two or three axis.

> A display device - either a CRT (usually) or hardcopy unit.

> Some means of interacting with the display - a keyboard to enter commands, a light pen or tablet to pick from a menu or point to an object, or a joystick.

The real fun begins as the data is being entered. The way data is stored is a major consideration. Given a set of coordinates, you must store them as well as their relationship to other locations on the object.

The programming for these packages can be complex and
expensive - particularly if the scaling (varying the
size of the object), translation (moving the object
from one spot to another), and rotation (turning the
object) are done in software - the alternative is to
do it in hardware, the software cost goes down, but
then you pay for it in iron (and glass). The simplest
case is the realization of an object - without hidden
line removal - in three dimensions (two dimensional
representations are not of much interest as a general
case).

Here we need to be able to describe the object, in
three dimensional space, pick a point of view and then
determine what that object would look like when
projected onto a viewing surface imposed between the
observer and the object - try sketching the chair you
are sitting on when viewed from any angle. Now
consider the added complication of having the observer
- you - be able to take any reference point, including
a spot inside part of the chair. Don't forget that
your viewing screen isn't infinite in extent, it has
physical limitations and as a result, the part of the
image that extends beyond the screen must be clipped.

To lessen the confusion of having unnecessary lines -
parts of the image that might be blocked from view by
a solid part of the chair, introduce hidden line
elimination.

The development of software and hardware for these
applications is expensive even for a limited library.
A complete applications system would easily run in the
$100,000 end user purchase price.

1.2.1 Hardware Requirements -

The CRT displays are usually vector devices, usually
of high resolution (1024 x 1024) to provide clean
lines. Raster equipment with its lower resolution and
"jaggies" does not provide any advantage until we add
the complexity of the solid fill for surfaces, or a
range of colors. Hardcopy - frequently pen plotters -
is a normal requirement.

1.3   IMAGE PROCESSING

One key difference between this application area and the previous two is that the image is usually a representation of a real object. It might be a LANDSAT photograph or a dental x-ray that needs to be enhanced. The problem that is dealt with here is not necessarily the display of the image – though that factor is here – but rather the use of graphics to extract more information from the data.

Consider a satellite photograph taken with a multi-spectral camera. Rather than viewing a piece of geography as we might in a more conventional camera, the multi-spectral unit uses filters to record detail at four (as an example) wavelengths. A green filter might be used to detect vegetation, an orange for soil, another for water, etc. Each of these can be digitized so that the computer has four copies of the same image – four two-dimensional arrays, each element of which is a measure of the intensity of light for the appropriate wavelength. By ratioing the images from the orange and green filters, we can emphasize the vegetation or barren soil.

Dental x-ray images contain more information than might be apparent to the human eye. The loss of detail is due to the low contrast level. After digitizing the image with a sensitive densitometer, the intensities can be rescaled and displayed with greater detail than we would have seen in the original image.

Image Processing is a major growth area for graphics and the interpretation of images. The pattern recognition aspects are of much current interest. Lockheed has recently been running advertisements in magazines illustrating the problem of automatically detecting tanks in battle situations (adding the desirable feature of distinguishing them from us was also noted). Closer to home, we have the area of robotics in automated manufacturing and parts inspection. Western Electric some years ago reported on their efforts to find faulty drill holes in fabricated parts automatically. During the recent (March, 1982) Corporate Research open house, they showed work on the problem of parts inspection.

The list goes on including the "deblurring" of photographs and their enhancement. A recent article

in Scientific American ("Image Processing by
Computer", Cannon and Hunt, October 1981) provides a
good overview.

1.3.1 Hardware Requirements -

The display hardware requires a high resolution unit
(512 x 512 can be acceptable for some use, others may
require 1024 x 1024) capable of at least 16 gray
shades or colors (the human eye can distinguish 64
shades of gray). Some form of hardcopy is needed.
The most appropriate is a photographic device (Dunn
Instruments or Matrix - about $10,000).   A "pick"
device - a light pen, cursor, or tablet - is sometimes
desirable to indicate special regions of interest.

Beyond the graphics equipment we also need to look at
the computer and storage sub-systems.   Digitized
images take up a lot of disk space, a single 1024 x
1024 x 8 frame is 1024K bytes. We also need to be
concerned about the movement of that data from storage
to the display. High resolution real-time animation
can put a substantial load on a CPU, so a high
bandwidth is important (we must be careful to
distinguish between real-time animation and simply
animation, in the first case high throughput is
needed, in the second, data can be recorded on film at
a slow speed and played back at normal speeds).

1.4 DOCUMENT PREPARATION

Throughout this document (prepared using RUNOFF, a
text processing package) you will see examples of text
formatting; automatic generation of
table-of-contents, indented and bulleted lists, bold
type and other features.   You will also see some
rather simple minded examples of illustrations -
graphics - mixed on the same page as text, as well as
more sophisticated graphics on separate pages.

While a combination of the VAX EDT editor or the

PDP-11 version - KED - and RUNOFF make a reasonable
text editing and production facility (once you get
used to the editor and the command structure of
RUNOFF), it does not allow the integration of
formatted text and pictures beyond what you see here.
The goal of such a product would be the preparation of
a document with the end result looking like a book.
There graphics appropriate to the subject matter are
found together on the same page rather than a page or
two away.

1.4.1 Hardware Requirements -

The key component of the type of system is the
printer.   It must be letter quality and still be able
to function as a graphics device. Ideally it would be
able to function like letter quality printer on the
Word Processing systems, able to take either tractor
feed paper or single sheets (the roll form paper is
awkward to separate into sheets and there is a high
likelihood of mechanical damage to the paper during
separation). The production of graphics could be
either through a printer that was "smart" enough to
interpret the graphics commands or have it function as
a screen copier.

A full page CRT display would be the ideal choice as a
preview and production device. Since the entire page
would be available as a drawing medium - giving a
one-to-one image on the printer - the page layout
could be composed on the screen, treating the text
material as graphics and then doing a transfer of the
bit plane to the printer.

COMPUTER GRAPHICS HARDWARE AND SOFTWARE

2.0 HARDWARE

2.1 Graphics Displays

Rather than to give a detailed tutorial on graphics
hardware, the intent of this section is to give you a
overview of the type of display hardware available.

CRT displays can be broken down into two categories:
vector and raster technologies. A vector is a line
drawn from some current position to a new one. This
type of display is also referred to as a "random scan"
device since the pattern of painting the screen
depends on the figure drawn. With raster systems, the
vector is interpreted as the dots (their position)
needed to draw it. In addition, raster devices paint
the screen in a ordered fashion, regardless of the
resulting image.

2.1.1 Vector Displays -

The basic concept of a vector display is something
most of us have be used to since we were children -
basically a connect-the-dots approach to drawing
pictures. The typical resulting figures from this
type of display are line drawings rather than filled
in areas. The lines are sharp, owing the the method
of drawing.

Consider a piece of graph paper (the common quadrangle
will do nicely) or the figure below. The
intersections of the lines are the addressable points
on the display surface. The number of lines - the
resolution - (1024 x 1024 is common) is governed by
the design of the hardware, specifically the
digital-to-analog converter (D/A) used (10 bits for
1024 points). The D/As (2 are used, one each for the
horizontal and vertical positioning) are used to

generate a voltage that is in turn applied to pairs of
deflection plates (one set for the horizontal axis and
one for the vertical axis). That voltage causes an
electron beam to be deflected from its normal straight
line path to another point on the screen. Just as the
graph paper is a continuous writing surface, so is the
display screen (it is evenly coated with phosphor), so
as the beam moves from one addressable point to
another, it leaves a straight line track on the
screen. It is the same as choosing two points of
intersection on the paper and joining them with a
line; the line is continuous between the points.
Repositioning the beam without drawing can be done by
not intensifying it during it movement. These two
modes of operation give rise to the two fundamental
graphics commands - move (repositioning without
intensification) and draw (repositioning with
intensification).

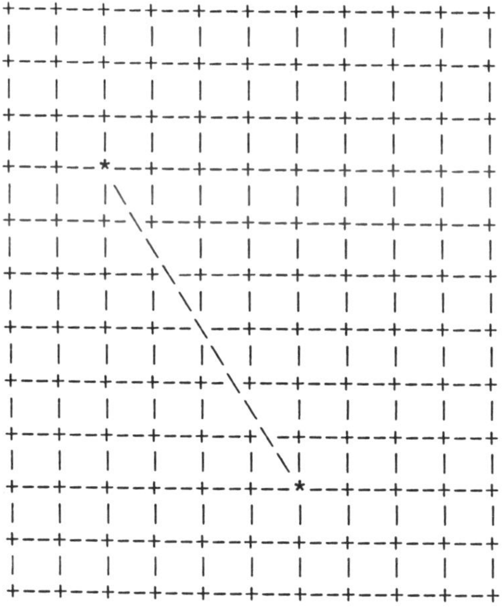

The * shows the
beginning and
end of the line
segment with the
\ joining them.

There are two basic versions of this technology:
storage tubes and vector refresh displays. Storage
tube technology (developed and owned by Tektronix)
allows the image to be "stored" on the screen. This

method permits a lot of lines to be drawn on the
screen, but does not allow the user to change part of
the image without redrawing the entire display which
can take up to a minute to do (in order to change it
the entire screen must be flooded with electrons and
then the image decays - the cause of the green flash
on Tektronix tubes). The contrast is low and only
monochrome images can be produced.

Vector refresh displays on the other hand can be
updated without having to reproduce the entire
picture. The figure on the screen is completely
redrawn - refreshed - every 60th of a second (this is
an automatic process) and as a result, and changes are
quickly seen (there is no flash to erase the screen).
The problem that this introduces is this: if the
image cannot be redrawn (due to having too many lines)
in that time period, an objectionable flickering (see
below) of the image will result.

The primary advantages of vector CRTs are the ability
to produce sharp clean images. The drawbacks are the
limited colors available (monochrome usually,
expensive beam penetration units can produce several)
and the fact that the user has to choose between a
static display for a large number of lines or contend
with flicker on a display which can be updated.

2.1.2 Raster Displays -
Raster technology provides a different approach to CRT
graphics. Raster devices have one thing in common
with vector refresh displays - they must be updated 60
times a second. But rather than moving a beam between

Flicker - when an electron beam strikes the surface of
a CRT a chemical compound [or mixture of chemical]
coating that surface - referred to as a phosphor - is
excited and emits light for a short period of time.
The decay of that light is not instantaneous but takes
some time, referred to as the persistence of the
phosphor; there are a number of different types of
phosphors with different persistences and colors.
When an image is drawn on the surface it is done with
a moving electron beam which can only be in one place
at a time. If the beam gets back to a spot that
should be excited before the phosphor dims
substantially, there is no problem, if it takes too
long, the light will be emitted in pulses - visible to
us - and the image will be seen to flicker.

random points, raster units scan the screen in a fixed
pattern, beginning at the upper left corner and moving
along a horizontal line. When that line is finished,
it moves to the left of the next line and refreshes
it, and continues this process until the entire
display has been completed. This approach introduces
a significant step between our "natural" approach to
drawing – vectors – and the resulting image. Any line
we want to draw has to be transformed from a vector to
something that the raster process can work with – that
transformation is called a scan conversion.

Instead of storing the location of successive
coordinates as might be done in vector units, raster
displays store the scan converted image in local
(within the terminal) memory. It is the amount of
memory in the terminal that limits the accuracy of the
resulting image. In the simplest case – monochrome
black-and-white – the screen is broken down into a two
dimensional array of pixels (picture elements). In
very high resolution displays each dot on the screen
(the tube technology is the same as television) could
be a pixel, but most raster displays are of low to
medium resolution (190 x 240 for low, 240 x 768
[VT125) for medium) and here a group of dots would
define a pixel. With very low resolution devices,
such as the Atari video game unit, the pixels can be
seen as filled boxes. Each pixel would be represented
as a single bit of memory. If we want to move into
color, we need additional bits of memory for each
pixel so that the color can be described. The VT125
for example, support four colors per pixel, so two
bits per pixel are required. The amount of memory
that the terminal contains puts a limit on the
sharpness of lines. One characteristic of raster
displays is something called the "jaggies" which
result when a line is drawn.

Its not difficult to see why the jaggies occur. To
begin with, lets go back to that piece of graph paper.
This time we will pay attention to the boxes on the
paper rather than the lines. Each box – or pixel – is
the addressable element in the display. If we want to
draw a line on the paper in the manner of the raster
display, pick boxes for the endpoints (a diagonal
shows the problem best) and draw a straight line
connecting the centers of those boxes. Now shade in

any box that the line passes through - the stair-step
pattern shows the jaggies. The higher the resolution
(the smaller the boxes) the less apparent the problem,
but its still there.

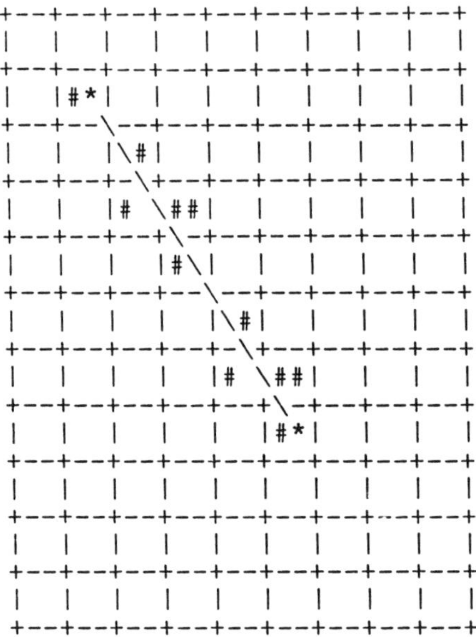

The * are the
beginning and end
of the line segment,
the \ is the desired
line and the # are
used to show the
filled boxes. Each
box represents a
single pixel.

There are some problems associated with going to
higher resolution.   Two   in   particular, the cost of
additional memory - we have to store the images
locally and each added pixel will cost one or more
bits depending on the number of bit planes, and the
phosphor coating on the display itself. The phosphor
in the VT100 for example is not a long persistence
phosphor (if it were the text would appear to smear in
the smooth scroll mode - this can be seen by using
smooth scroll for full lines of text in a dark room).
In order to paint the screen with   double   the
resolution of the VT125, we would have to go to an
interlace mode in which the odd numbered lines are
refreshed on one pass and the even lines on the next.
With a short persistence phosphor, that would lead to
a flickering display.   The only way to overcome it is
to change the bottle - the CRT tube - and that would
eliminate field upgrades [Retrographics - a company

that rebuilds VT100 for higher resolution graphics
(480 lines vertical) does change the bottle and uses a
longer persistence green phosphor].

Raster units do have some distinct advantages over
vector devices:

o   Low cost - a typical vector device might begin at
    $7-8,000 (the VT-11 began a $11,000 some year
    ago),

o   Color - depending on what you are willing to pay,
    you can purchase 4 or 8 color units for less than
    $4000 and get terminals with 512 x 512 pixel
    resolution and a choice of 256 out of a palette of
    16 million (AED 512) for $15-18,000. Some limited
    color is available on vector unit (using beam
    penetration techniques for example, [Evans &
    Sutherland]), but at high cost.

2.1.3 Raster Color -

Color is another factor in raster displays. Using
vector systems we were limited to a single color
against a dark background, here we have more
flexibility. By adding more memory in the form of
additional planes, we can store more information about
each pixel. With only one plane, we can tell if that
pixel should be illuminated or not, giving a
monochrome black-and-white display. The second plane

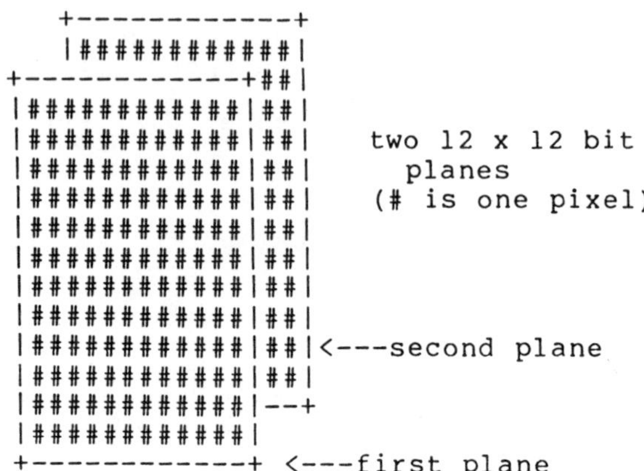

can be viewed as being "stacked" behind the first, so
that each pixel can now be viewed as having two bits
of description. The combination of bits gives us four
choices (0, 1, 2, or 3) for a color for that pixel.
The VT125 has a table which allows us to describe four
colors and reference them by the numbers 0 through 3.
By adding additional bit planes, we can have more bits
per pixel with which to choose colors, three bit
planes would give us eight colors.

2.1.4 Connections To The Host -

With both vector and raster technologies, there are
also the questions of tightly couples vs loosely
coupled devices and graphics protocols.

The VT-11, VS60, VS-11, and the VSV-11 are some
examples of past and current DEC graphics devices
which fall into the category of closely coupled
devices.   In all of these, the CPU and the controller
share a segment of main memory. That memory is used
by the users program to store graphics instructions
(they could be entered directly by the user or result
from calls to a subroutine library) which are read by
the graphics display controller.   These   devices
execute the instructions in that shared memory each
time the display is refreshed.   That close-coupling
permits changes to the screen to occur rapidly, as
might be required in real-time animation or when a

model of an object is rotated as it is viewed. For example if we wanted to watch a box tumble in space or see the three dimensional structure of a molecule as it is rotated, this would be the type of linking between the CPU and display desired because of its ability to make rapid changes to the screen. Imaging is another area where having closely coupled devices is important.

Take the GAMMA-11 system (a system used for studying images from gamma cameras) for example. One of the types of studies that is possible on that product is the ability to show successive pictures of an the distribution of a radio-isotope within an organ as a function of time. Through the use of a closely-coupled display, the sequence of frames could be seen as a "movie". Without rapid updating of the screen, the effect would not be possible.

These types of devices - also referred to as a display list device or frame buffer device - are high performance and expensive.

Terminals tied to a computer through RS232C or 20 mA connections are an example of loosely coupled devices. The serial ASCII transmission of data is adequate for applications where dynamics or animation are not a factor (the VT125 and GIGI can be used to a only very limited extent here). This approach has been used in both classes of equipment, our own terminals for example in raster work and Tektronix or IMLAC terminals in vector systems.

Closely coupled and loosely coupled devices usually have different types of graphics instruction sets. In the former case, the instruction set would be likened to machine language in a computer, they are concise and stored in a ready-to-execute format so that the processor can work with a higher level of efficiency appropriate to a higher performance device. Loosely coupled devices may or may not have an efficient format. ReGIS is one example of an instruction set that must be interpreted before any drawing can occur, much like an interpretive Basic. A display list device my have an emulator for a ReGIS type instruction set to take advantage of a broader range of software (RGL or Datatrieve graphics on a VS-11 class unit).

Graphics protocols - the method of describing and communicating graphics information between the

computer and the display device - is another area of
concern.   It bridges hardware and software domains,
since it is usually embodied in firmware within the
terminal (ReGIS is an example). Before we get too
involved in this area, lets finish some points on user
hardware requirements and deal with protocols in the
context of software.

2.1.5 Graphics Hardcopy -

The display of images on a graphics display is usually
only the beginning of the users requirements for
graphics processing. He is in the same position that
we would be in describing a television program to
someone who hasn't seen it - verbal descriptions are
nice, but they don't compare with seeing the actual
image or program. Some form of hardcopy is frequently
necessary, some rendering of the CRT image that used
in a presentation or publication. Here we can find a
wide range of devices depending on our requirements
and ability to pay.

o   Screen copiers are available (Tektronix and
    Honeywell produce these as do others); these
    produce an image through the use of the composite
    video signal from the terminal. The images are
    not always permanent and of low contrast, although
    some low cost ink-jet units might change this
    picture.

o   Photographic systems, such as Dunn or Matrix
    Instruments (or a 35 mm camera) can produce color
    copies of the screen. Here you get what you pay
    for, a 35 mm photo taken from a screen might be
    satisfactory for some applications but if you want
    good color fidelity, no distortion due to screen
    curvature or camera optics, the high priced
    ($10,000) units may be required.

o   Raster printers may product workable black and
    white images, but only paper media.

o   Ink jet printers can be used to produce good
    quality prints in color, but usually at a high
    cost.

o   Pen plotters can be used for to produce hardcopy
    graphics on both paper and transparent media with
    color if the unit has multiple pens. These can be
    had at low cost, but may be a problem if the
    graphics protocol differs from the CRT -
    duplication of software.

2.2 Hardware - User Interaction

Viewing the graphics output of a device - either a CRT
display or a hardcopy unit - is only one side of the
users interaction with the graphics system. In some
cases it is all that is needed, but the are many
instances in which direct interaction is desirable or
mandatory. In this section we will take a look at
some of the devices used.

2.2.1 Light Pens -

The term "light pen" is a misnomer, rather than
writing with light, the device - a long narrow tube -
detects light. The way a light pen functions is
really quite simple. When an image is being drawn on
a CRT a single electron beam is used. That beam can
only be in one place at a time (this is true or both
vector and raster technology except in the case of
storage tube). If we place the light pen on the
screen - it must be a place where the tube is
illuminated - the pen will generate an interrupt to
the graphics hardware when the electron beam crosses
the pens field of view and causes the phosphor to emit
light. When that occurs, the current position is read
by the hardware and returned to the program. Storage
tubes the Tektronix 4010 or 4014 for example, have a
built in cursor that can be moved around to serve the
same function. The reason that a light pen doesn't
work in this case is that the entire image is being
maintained by a continuous flood of electrons rather
than a single beam.

Light pens are referred to as a "pick" device in the

proposed SIGGRAPH graphics standard be cause they can
be used to indicate - pick - a point of interest.
Both the VT-11 and the VS-60 incorporated light pens.

2.2.2 Joysticks -

Joysticks are among the more common of the graphics
input devices, with the highest percentage used in
video games [during a recent trip to a shopping mall
to research this point, it was found that most home
video games and arcade games employed this means of
human interaction]. The principal attraction for this
approach is that it is inexpensive and affords a
natural eye-hand coordinated input to the system [the
games that relied on buttons were not as easily
mastered for example - unless you are an well
coordinated 10 year old]. Visual feedback is usually
by means of a cursor on the screen. BYTE magazine
(January 1982) ran an article on the construction of
one for TRS-80 model I or III.

Like the light pen, the joystick is a two-dimensional
pick device.   The Gamma-11 system used a joystick to
indicate regions of interest for medical images.

2.2.3 Track Ball -

The track ball is not as common a device as are the
others in this section. Physically it is a ball that
is mounted and can be rotated in any direction [the
arcade version of Missile-Command uses this device].
As the ball rotates - the direction of rotation
indicates the desired direction of movement - the
position of a tracking device can be read by the
computer (using 8 or 10 bit A/Ds) and a cursors
position updated to give visual feedback. Part of the
problem with this type of device is its size, the ball
may be four to five inches in diameter in order to be
easily manipulated.

2.2.4   The Mouse –

The mouse in essence is a track ball mounted upside-down. It is a handheld device that can be moved around on a table or desk top to indicate relative motion. Move the mouse to the left and the visual response – the movement of a cursor – moves to the left. Because of its size, it is convenient to work with and relies on a natural eye-hand coordination. The is no absolute positioning with the device (it can be lifted and put down in a different spot without the machine being aware of it). The PERQ system uses this device to interact with the system.

2.2.5   Tablet –

The tablet is a small digitizer, perhaps fourteen inches on a side. Using either a stylus or cursor, it can feed back coordinates (within a fixed frame of reference, thus it is an absolute positioning device) to the machine which can be used to select a point of interest or enter data by tracing a curve. Those coordinates must be interpreted by software in terms of the screen coordinate system since is it unlikely that they will match.

There are a number of manufactures of these devices: Talos, Summagraphics, Gteco, and Houston Instruments are but four. The GIGI firmware has the logic to handle the Summagraphics Bitpad built in, and uses that device in its locator mode (movement of the cursor on the table is echoed as a corresponding movement of a cursor on the screen).

2.2.6   Touch Sensitive Screens –

Touch sensitive screens are another mechanism for indicating a choice of items on the screen. It is a course device, with the limit to its resolution governed by the size of a finger, rather than electro-mechanical constraints in the previous devices. The method of interaction is to point to and/or touch the object or menu item on the screen.

There are two common technologies for providing this

capability.   One consists of two layers of membranes.
When touched the capacitance between these layers
changes and the position of that change can be read by
the hardware.

The second approach involves rows of LEDs and
detectors, one each for each axis, when the path of a
ray of light is broken, the position can be fed back
to the machine.

Besides the low resolution, the prime drawback to
these devices is operator fatigue. Occasional use
would not be a problem but continuous or frequent use
is.

2.2.7 Final Note -

All of the above points are concerned with the
hardware associated with a persons interaction with
the graphics displayed for him, the way in which a
choice is made rather than how that choice is
presented. It is the latter point that can make an
users experience with a system range between a
pleasure and a pain. Type written commands may be
appropriate for a touch typist given a large choice of
options, but a burden for the seek-and-ye-shall-find
(also known as hunt-and-peck) variety.

"Menu driven" software is useful when a limited number
of commands are available or if they can be grouped
into sets of menus. The problem encountered here is
that multiple menus eventually wind up in a tree
structure and getting from one branch to another is
not always easy (or may be too time consuming).
DECgraph - a data plotting package under development -
uses icons (drawings representing a function) in a
menu format to choose between options.

Special function keys, such as those in some word
processing systems are a hybrid of both of the above.
The keys can either represent commands or a form of
interactive menu depending on your point of view. In
some of the Hewlett-Packard terminals the split
between functions is clearer. Keys located under the
face of the screen can be used for menu selection,
while others can be programmed to enter preset
commands (when a key is struck, it causes a user
defined character string to be sent to the system).

The HP approach is appealing because of its flexibility.

There probably isn't any single best approach. One must look at the application and decide which one or which combination of several might be appropriate. From the standpoint of hardware design, we need to offer the broadest range of choices possible.

The point of this discussion is to let you appreciate that the design of a graphics system, or application, needs to involve more than hardware, but has to take into account how an individual is going to use it.

2.3 Graphics Software

In terms of structure, graphics software is analogous to the levels of software that exist in operating system/programming language environment. Higher and higher levels of drivers, system utilities, and languages exist in the latter with equivalent entities in the former. A disk drive controller has a command register which controls the positioning of the heads and data transfer, similarly the graphics display will have its set of instructions. To gain device independence, we communicate to disk controllers through software device driver or handlers, the current state of technology in graphics suggest the same approach; a common set of commands is dispatched to the handlers and they take hardware differences into account. The file control system can be likened to a graphics library and so on.

At the lowest level, we have the graphics instruction set (also referred to as a graphics protocol) for a given graphics display. As noted above, this is the means of telling the device what we want done; the interpretation and execution of those commands take place within the graphics terminal or subsystem. ReGIS is the graphics protocol for the VT125 and GIGI. Every graphics device has one, although each vendor may implement one or more for his equipment. The complexity of the protocols vary widely. Some may have only a few commands, to do a move and draw (discussed above in the section on vector displays),

plus an erase command, others may have dozens of
commands. The difference between the two is usually
reflected in the cost of the display and the amount of
software   that   the   user need write to get a job done
(the richer the command set, the easier it is to get
things done).   To   give an example some plotters can
print text simply by being told what characters to
print out, others have to be told how to form each
character.

There are no formal standards for these commands sets,
each manufacturer may create and implement his own or
emulate the instruction set of another manufacturer if
he   feel   that   it is beneficial. Many have chosen to
use   the   instruction   set   for   the   Tektronix 4010.
Although very limited in it command set, the 4010 has
proven to be a popular terminal (see below) .   That
popularity   and   the   availability   of   a   wealth   of
software for it has fostered the emulation of the
device by a large number of vendors of display
hardware who interest is to sell hardware and not fund
major   software   development   to   compete   with a well
known library. This protocol has been emulated by
Digital   Engineering,   Hewlett-Packard,   Advanced
Electronic Design, Princeton Electronic   Products,
Vector Automation as well as others to give them ready
access to software. The reason for it is simple:
software libraries provide the programmers interface
to the graphics hardware and software - particularly
for graphics - is expensive to produce. It is easier
to work with established packages rather than go
head-to-head with them.   The terminal replacement
market, particularly for Tektronix replacements, is a
profitable one.

ReGIS is the protocol for the currently available DEC
graphics terminals.   It differs from the bit encode
scheme of the 4010 and 4014 devices   as   well   as   the

There are three Tektronix terminals that are emulated,
the   4010,   4014   and   4027.   The   4010   is   a   high
resolution terminal with cross hair cursors   for
graphics   input,   the 4014 adds a hardware line
patterns, choice of character sizes and user definable
character sets with local storage for commonly used
structures. The 4027 is a color raster terminal with
64 colors, 16 character fonts, as well as polygon and
vector commands.

VT105 in that it is a richer command set and is a
simple ASCII stream, it can be easily read with out
resorting to decoding the bit patterns of characters.
Graphics figures can be stored as simple sequential
files, edited with a standard editor and transported
between system (ours or other vendors) without
worrying about the oddities of binary file structures
and data being interpreted as control codes .

Given the variety of graphics devices and instruction
sets, a programmer may find that a package that works
well on one vendors hardware will not work at all on
anothers. Because of this, and the needs of differing
applications, there are a lot of graphics software
packages in the world, most of which have little
regard for one anothers existence (program
transportability between systems is nil).

As a result of all this a few things have occurred.
Several packages have become preferred over others:

o   Plot-10: A package written originally for
    Tektronix equipment, and functioning for any
    terminal that has 4010 emulation. The cost of
    this system can run from as low as a few thousand
    dollars for some minimal capability to thirty
    thousand for a full blown software system. The
    features of this package include software
    libraries as well as interactive graphics and
    drawing routines.

o   GCS: This is a package originally written by the
    U. S. Army in Vicksburg. Though not an
    up-to-date system its prime attraction is that it
    is free, and has all the support that the
    purchaser can provide. This package has gotten
    around the problem of differing protocols on
    different hardware by providing an interface layer
    between the subroutine library and device. 4010
    emulators are common and Sandia has produced one
    for the VT125.

o   DISSPLA and TEL-A-GRAPH: These are commercially
    available packages, both highly regarded, sold by
    ISSCO in California. The price tag is about
    $40,000 for the full software package and it will
    run on a variety of computer systems − including
    VAX/VMS − and terminals using the same approach as
    GCS for dealing with different devices. The
    output of these packages is impressive, some
    examples follow (figures 6a and 6b):

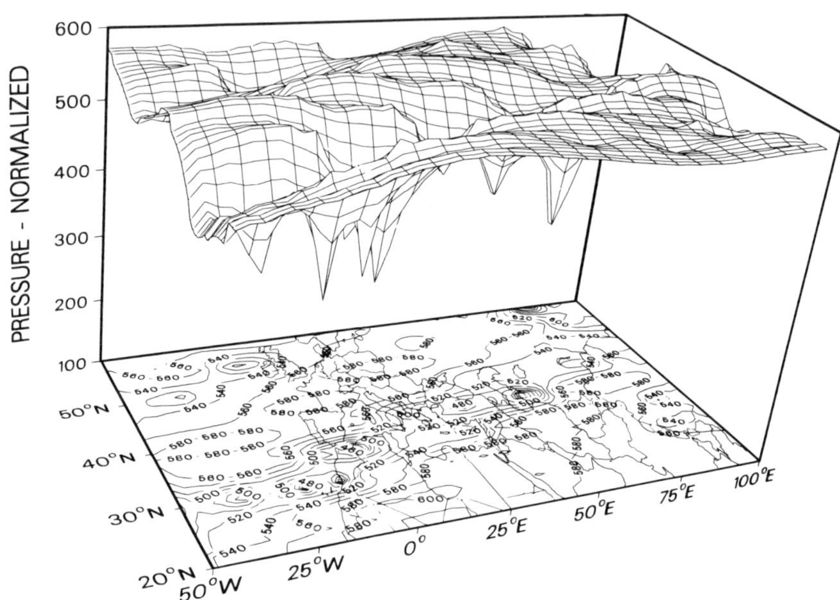

Figure 6a.   Accumulated Pressure Changes - 1980.   Reproduced with permission of ISSCO.

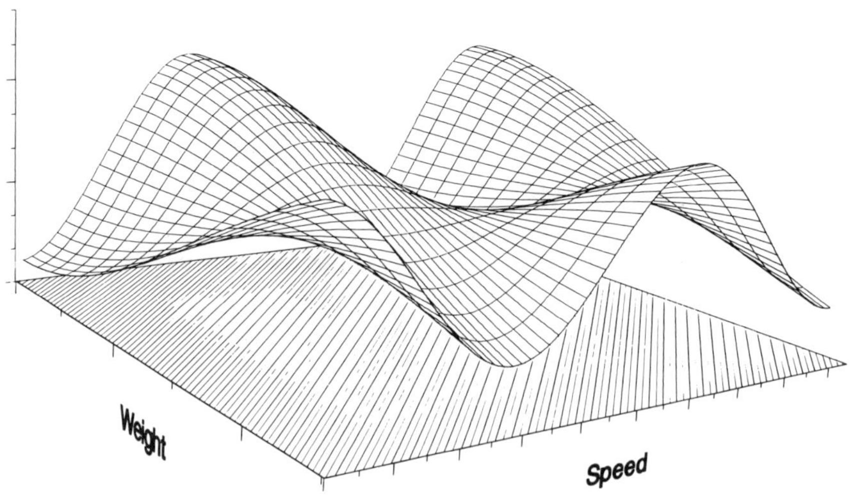

Figure 6b.   Vibration Analysis.   Reproduced with permission of ISSCO.

Beyond this has come a significant development — a movement toward standardization of graphics software. Of the efforts underway, two are particularly significant: the SIGGRAPH CORE effort (in conjunction with the ANSI X3H3 committee) and the GKS (Graphics Kernel System) from Europe. The essential problem is a uniform interface between the applications programmer and the graphics hardware he is using. The worst case situation is one in which a different graphics package has to be written for each device giving no portability at all between systems. Here the hardware dependences are seen directly by the programmer as shown in the figure below.

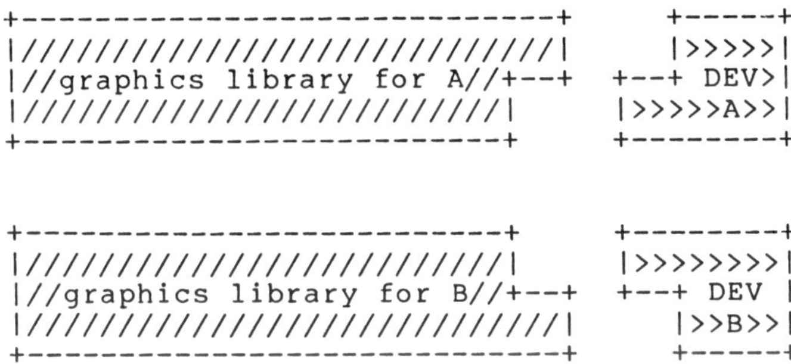

At the next level, the subroutine library that the user works with remains functionally constant, but its internal structure will directly take into account differences in hardware. This allows some portability, but puts the burden of support for different devices on the system/library manager.

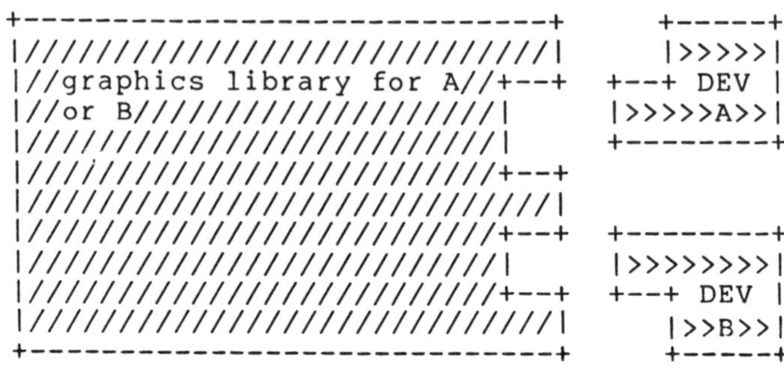

```
+--------------------------------+      +-----+
|////////////////////////////////|      |>>>>>|
|//graphics library for A//+--+    +--+ DEV |
|//or B//////////////////////////|      |>>>>>A>>|
|////////////////////////////////|      +--------+
|///////////////////////////////+--+
|////////////////////////////////|
|///////////////////////////////+--+   +--------+
|////////////////////////////////|      |>>>>>>>>>|
|///////////////////////////////+--+   +--+ DEV |
|////////////////////////////////|      |>>B>>|
+--------------------------------+      +-----+
```

At a higher level the graphics library can remain
constant, and hardware differences can be taken into
account at the device handler - the lowest level
before talking directly to the hardware, as shown
below. The library would dispatch standard messages
to the handler and it is that items responsibility to
implement them. This allows a high degree of
portability and the easiest integration of mixtures of
devices and it is at this structure that the proposed
standards are aimed. The proposed CORE standard in
particular deals with this level and the segmentation
of software. As noted much earlier, a software system
can exist at a primitive level and be build upon,
adding layers of software as the applications dictate.
This layering is directly addressed by the CORE,
defining what capabilities will exist at each level
(both graphics input and output) the functions of
various subroutine and their relation to each other -
it is a functional specification for a layered
software product.

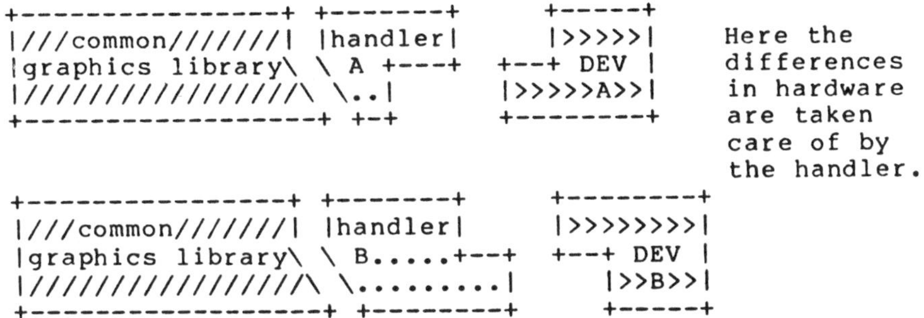

The problems with these standards is that they are
still years away. Thus packages that state that they
are CORE "compatible" rather them fully CORE compliant
- or compliant to a particular level - exist because
the CORE is a moving target, that movement being
forced by a need for completeness and rapidly changing
technology (the 1979 specification has raster graphics
as an appendix - it is now an major consideration).

If all this looks to you like a very messy situation,
then all this verbiage has had its point. But take
heart, their appears to be some light ahead and its
not an on-coming train. The bottom line for us is
that we must make sure that our software is consistent
across operating systems and hardware. On more than
one instance we have been told by customers that "we
don't care what standard you set, set one and use
across operating systems and hardware".

How does the software relate to the applications noted
above? Lets consider the diagram below (figures 7a
and 7b) to illustrate the layering of software (the
diagram was originally presented by Tom McIntyre,
Central Engineering, Digital Equipment Corporation).

At the lower left corner we have the terminals with
some on board intelligence. The software does not
exist and the user must not only worry about his
application, but how to get the hardware to do what he
wants done. The software is device specific. As we
move across the bottom, the hardware gets "smarter"
and takes on more of the burden. As we move up, the
software get more powerful, and further removes the
application programmer from the hardware, he worries

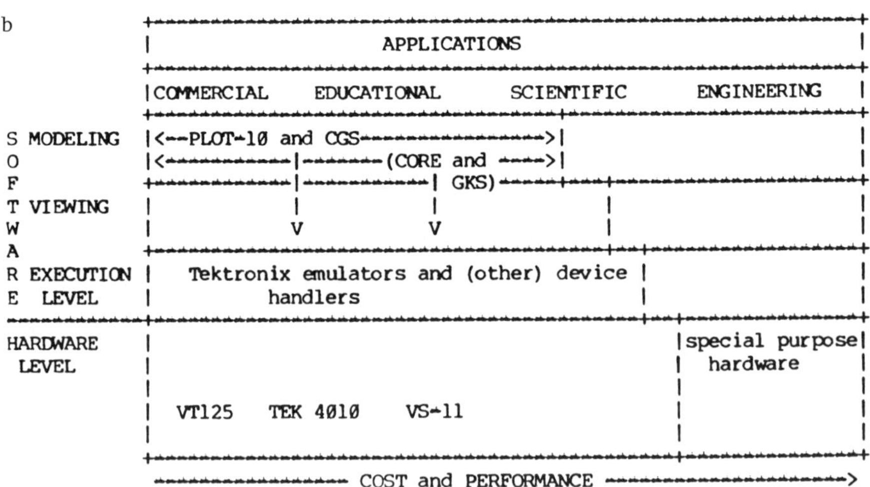

Figures 7a and 7b. Graphics Product Environment.

about his application requirements and thinks in terms
of the graphics figures he is drawing ("I want a box
here", rather than "how do I draw a box on this
terminal"). The characteristics of a particular
terminal (or its limitations) are taken into account
by the software. At the top, we concern our selves
with the needs of a particular problem, using a data
plotting package, rather then trying to figure out how
to label the axis. Lets take a look at the same
diagram with some of the points on software noted
earlier added.

The International Standards Organization, of which
ANSI is a member, has approved the Graphics Kernel
Package as its graphics "standard" of choice. This in
effect, makes the GKS approach the ANSI standard.
While this would appear to settle the issue of CORE vs
GKS, all it really does is provide a modification of
direction. There are a lot of graphics software
systems that are built on the CORE approach, which
represent significant financial and manpower
investments, and they are not likely to wilt away.
The closing salvo of the standard battle is still a
long way off.

3.0 SUMMARY

Through the course of this article, I have tried to
present an overview of the scope of graphics and an
idea of what can be done, and what it takes to do it.
In a field as fast moving as this, everything said was
out of date as soon as I finished typing. To give you
some idea, for a few hundred dollars ($100 - $200) you
can purchase a digitizing tablet and some software
that allows a popular microprocessor system to become
a free-form drawing package.

Because of the amount of information that can be
presented and the increased clarity gained from
graphics, it is going to find an increasingly
important role in the laboratory. Compared to dealing
with reams of numbers, it will be a welcome relief.
The software and hardware is becoming less costly and
easier to use. It will not only change the way the
data in the laboratory is used, but make it more
interesting and allow us to extract more useful
information, faster.

Trademark and Product Acknowledgements

o   4010 and 4014 are terminal model numbers for
    products of Tektronix Inc.

o   Plot-10 is a trademark of Tektronix Inc.

o   DECgraph, Gamma-11, GIGI, PDP-11, ReGIS, VSV-11,
    VT105, VT100, VT125, VT-11, VS-60, and VAX are
    products of Digital Equipment Corp., Maynard,
    Mass.

o   PDP, VAX, and VT are trademarks of Digital
    Equipment Corp., Maynard, Mass.

o   TRS-80 is a trademark of Tandy Corporation

o   DISSPLA and TEL-A-GRAPH are trademarks of ISSCO
    Graphics Software

o   AED is a trademark of AED, Inc.

o   Missile-Command is a trademark of Atari, Inc.

o   PERQ is a trademark of PERQ Systems Corporation.

RECEIVED July 31, 1984

Chemists and Computers in the Corps of Engineers

RICHARD E. ENRIONE

U.S. Army Corps of Engineers, Ohio River Division, P.O. Box 1159, Cincinnati, OH 54201

The chemists in the U.S. Army Corps of Engineers, Ohio River Division perform a variety of computer assisted tasks in assessing water quality and providing input to water management decisions. These include developing and running sophisticated water quality models; evaluating water quality data; and operating an automated water quality laboratory.

The U.S. Army Corps of Engineers has a relatively small number of chemists. They are located either in research laboratories or in the district and division offices. This paper will be limited to the Ohio River Division headquartered in Cincinnati and the four district offices in Huntington, Louisville, Nashville, and Pittsburgh.

There are approximately 75 multipurpose storage reservoirs and another 75 river lock and dam structures which are operated by the division. In their operation, the highest priority is given to safety of the structures followed closely by either flood control or river navigation. Depending on the particular project, several competing purposes govern most of the day-to-day operations. These include: hydropower, recreation, water supply, water quality, and minimum flow releases. The important point is that water control decisions have an impact on water quality. It is this aspect which requires chemical expertise.

There are three general ways in which water quality considerations impact on water management decisions. These are the long-term development of, or modification of operating guidelines; an intermediate term tracking of the effectiveness of the guidelines; and the real time or quasi real time monitoring of situations which have the potential for rapid change.

The guideline modification is exemplified by Bluestone Reservoir. In this case, proposals were made to modify the project operation in two different ways. First, change the release schedule to accommodate downstream whitewater rafting; second, increase the reservoir depth and add hydropower. To evaluate these, a mathematical model was used which incorporated the two dimensional hydrodynamic, thermal, chemical, and biological characteristics of the reservoir. The results of this modeling indicated that the scheduling of rafting releases would not aggravate the intermittent algae problem currently found in the lake and might help it slightly. On the other hand, the addition of hydropower would greatly increase the algae and also degrade the water quality of the reservoir releases to the detriment of the fishery below the dam. The modeling, particularly the computer generated graphics to illustrate for non-scientists the potential changes in algal growth, had two results. A modified release schedule for rafting will be implemented; and the hydropower addition is being delayed pending the results of studies aimed at reducing the nutrient load and a risk analysis of the impact of poor quality water on the downstream fishery.

The tracking of guidelines involves the routine sampling and analysis for a variety of chemical and biological constituents. At J. Percy Priest Reservoir, for example, there is a large historical data base for iron, manganese, ammonia, dissolved oxygen, etc. The monitoring in this case is to determine if the lake is changing in response to rapidly changing land use patterns in the watershed, and, if so, should the water management scheme be reevaluated. This effort results in a large number of samples for chemical analysis. There are five laboratories which perform most of these analyses. The four district laboratories perform the field sampling, biological and chorophyl analysis, and in some cases analysis for TOC, dissolved carbon, solids, alkalinity and acidity. The division laboratory performs typical water quality chemical analysis using, almost exclusively, mechanized/computerized equipment.

The real time data field data is collected hourly via a telephone network in the case of the Ohio River and through a GOES satellite and a downlink in Cincinnati for other locations. The satellite system, which consists of over 900 stations in the basin, was set up for flow forecasting. A provision was made to add water quality information and, as the need arises, appropriate monitors are installed at the required locations. Two examples of the uses of this information are: to change gate openings on the Ohio River Locks to maximize reaeration when the dissolved oxygen level get too low; and to monitor hourly fluctuations from petroleum brine discharges in the Blaine Creek watershed.

These efforts are carried out using a variety of computers with different primary purposes. Each district and the division have a water control minicomputer (typically a Harris 100) devoted mainly to hydrologic modeling on a real time basis, and the maintenance of appropriate on-line hydrologic data bases. Also, the real time water quality information is processed and analyzed on these machines. The division counterpart, in addition to these tasks, is used for the development of hydrologic models; the development and use of water quality models; and is the control point for the gathering, sorting and disseminating of real time data.

Each office also has a general purpose, engineering use mini (typically a Harris 500), which is used for the maintenance of the

district water quality data base and the running of various data
analysis and depiction programs. The division has a general
purpose Honeywell which is used principally for financial purposes
but augments the laboratory computer by generating most of the
management reports, performing some of the calculations and
quality control checks and acts as the control point for
disseminating of the data to the districts.

The district lab computers are applied differently in each of
the four districts. Among the uses are the storage retrieval and
analysis of biological data; the point of entry for field data;
data reduction of laboratory analytical results; and direct data
gathering from equipment such as spectrophotometers.

The division laboratory has two Wang VP2200 computers which
are interfaced to a variety of instruments. The nature of the
interface and the associated programming depend on the instruments
involved. In terms of computer usage, the instruments either
accumulate data which is then batch processed, or they require
continuous on-line support by the computer. For the ICAP, TOC,
and GC completed reports are sent to the Wang which require little
more than reformatting. For the Atomic Absorption, a data logger
sends in a sequence of numbers which requires some additional
processing. The electronic balance is operated in conjunction
with an interactive program for the analysis of solids. Six
channels of Technicon Auto-analizer are interfaced through a
fluidyne scanning A/D converter. Both of the last two instruments
place a considerable burden on the computer resources. In the
operation of any of thse instruments, the first step for the
chemist is to obtain a sample list from the computer; the last
step is to review the quality control data on the computer and
accept or reject all or part of the run. All computer operations
are menu driven question/answer sequences with the most common
responses available by default.

All of the efforts require a rapid interchange of information
and access to a wide variety of data bases, both external, such as
USGS's WATSTOR and EPA's STORET, and internal, containing
historical analytical results and reservoir hydraulic
information. This is accomplished by a network of computers tied
together by telephone lines and autodialing equipment, and is
shown in Figure 1. For example, on an hourly basis, a computer
calls the downlink to retrieve the latest set of satellite data.
This machine on a daily basis (or on demand) is automatically
called by the district water control computer to obtain the
relevant data. As part of the autodialing protocals, a user
working on a modeling problem in Cincinnati can access, and have
in his own fils in a few minutes, such information as current
water quality data from a district general purpose computer, flow
data from district water control machines, or historical chemical
data from STORET.

An integral part of the system are the backup procedures. To
deal with the problems of computer failure, loss of data base
integrity or communication failures, a combination of approaches
are used. The basic assumptions are: One days worth of
laboratory results on the lab computer are expendable in the sense
that a significant fraction is still in the memory of the

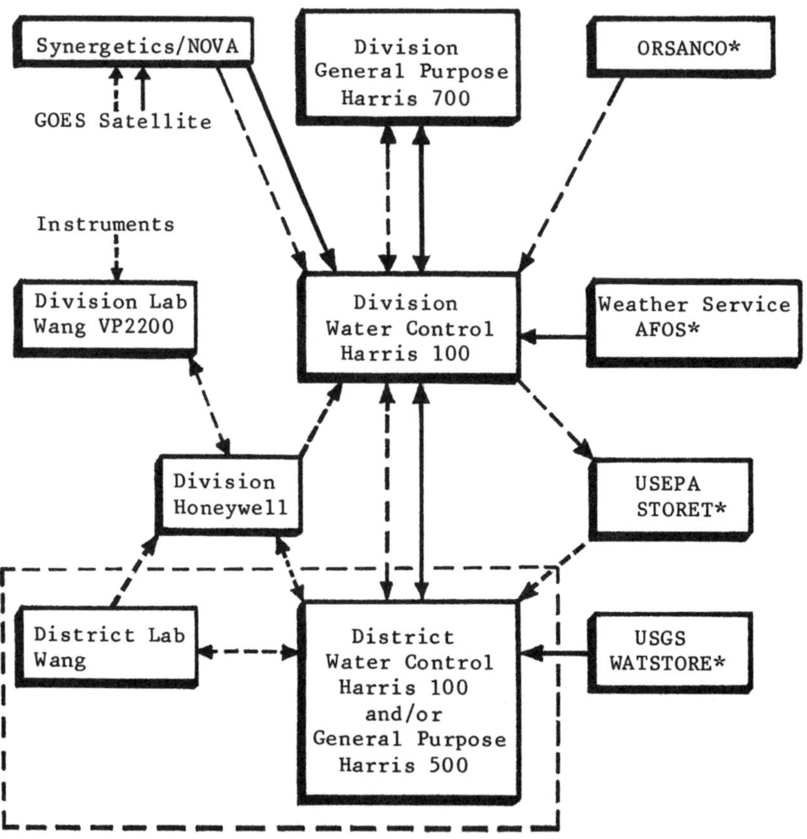

Figure 1--Schematic of network showing primary data paths.

→Water Control ---►Water Quality

*Equipment not operated by the Corps of Engineers.

The portion in the box is replicated in each of four districts.

instruments and the rest can be repeated. Real time hourly data for periods of 24-48 hours are expendable except during flood emergencies. All files on all computers except the laboratory Wangs are backed up on a daily basis with tape copies. The Lab computers send all results to a second computer every day. All appropriate results are sent to STORET quarterly. Within 24 hours, all current data is stored on computers at two or more sites (this is a direct result of the fact that the data users work on different machines than the data generators/data collectors). On a monthly basis, all programs and data files on all computers are copied to tape or disc and stored at a different site.

The computers and communication links which are critical at flood times have specifically designated backups where all the necessary data and programs are kept on a standby basis. For example, the Cincinnati water control Harris which is the key to disseminating real time data can have its function taken over by machine in Louisville; the local satellite downlink can be replaced by one in Mississippi.

In the lab, future expansion plans include the use of optical scanners for reading sample labels, operation of robots to relieve some of the manual operations and an artificial intelligence system to track quality control. In other areas, there will be an increase in the number of real time monitors, not necessarily because real time data is needed, but the cost can be small compared to sending out a field team. There will be some applications of direct monitoring by satellites such as LANDSAT D. Both of these will be incorporated into water quality models which will allow more intelligent choices of where to send a field team to collect samples for detailed analysis.

RECEIVED May 30, 1984

Computer Generation of Structure–Effect Relationships from Text Databases

RUDOLPH J. MARCUS[1]

Office of Naval Research, 1030 East Green Street, Pasadena, CA 91106

The use of text as data for clustering combines computer text processing, on the one hand, with a heuristic research methodology, on the other hand. In addition, the clustering of text data involves non-parametric hyperspaces, where definitions of the closeness of clusters are more difficult than they are in parametric hyperspaces used in the clustering of numerical data. Three different approaches to clustering in text material are described. The first one makes use of the format of two properties, such as structure and activity, on the same computer line or at least in the same file. This method assumes more importance with the addition of text searching to CAS ONLINE announced for this year. A second approach involves assigning vectors, such as +1, −1, or 0, to a property, an opposite property, or absence of a property, and then searching the data base for items with desired vectors in appropriate columns. A third approach is more quantitative and uses various projections of the hyperspace of properties onto one or more recognizable Cartesian axes. Examples are taken from a data base of chemical compounds and their medical uses, extracted from the Merck Index.

The work which my associates and I have done in the use of text as data by chemists and for chemists draws on two sources. The first of these is computer text processing. It is even more clear today than it was 13 years ago when this work was started that computers may not only be used for number crunching but for a number of other jobs in which letters and words as well as numbers can be used. Tasks such as extracting, sorting, reformatting, associating, and counting can be done not only with numeric information, but also with alphanumeric information. Whereas such operations may have been useful but esoteric 13 years ago, the tremendous use of word

[1] Current address: 605 Cavedale Road, Sonoma, CA 95476

processors and personal computers has made this an everyday event today. It is my contention that chemists are slow to use some of these alphanumeric, non-number-crunching techniques and it is interesting to speculate why chemists seem to be slower than those of many other disciplines, particularly psychology and sociology, to use these techniques. In the first place, chemists are used to dealing with formulae and with numbers more than with words. Chemical elements and compounds are more easily expressed as formulae than they are as words and certainly all of our figuring is done with numbers rather than with words, i. e., digitally rather than analog. However, as my coworkers and I have shown in a number of papers, formulae and the Geneva System names which these formulae express are alphanumeric information of the type that can be handled easily by the computer and which is therefore a fitting subject for the operations I was talking about earlier: extracting, sorting, reformatting, associating, and counting.

Not only has the microcomputer revolution changed the field enormously since this work was started, but also the resources which are available to chemists have increased tremendously. Chemical Abstracts and also the Merck Index, which we will be dealing with in this paper, are now available on tape. There has been the advent during the last few years of CAS ONLINE which is of tremendous help not only in searching bibliographic information but, as I intend to show in this paper, for actually associating information and therefore applying the inductive process to the data thus extracted from a data base. Methods of searching this large amount of machine-readable information have also been augmented during the last 13 years. Systems such as those operated by SDC and Lockheed Data as well as others have become very efficient in extracting information. It is a curiosity that the use of these systems is very much more developed in industry than it is in universities. Apparently in universities it is still cheaper to send a graduate student to the library instead of allowing intelligent use of these systems. I have dwelled so far on the extracting of information and I will deal in the rest of this talk with the other four processes I was mentioning earlier: sorting, reformatting, associating, and counting information.

Dr. Dessy, in his introductory talk at this symposium (1), argued very convincingly for greater use of these systems by chemists on the basis that anything less will leave the user overwhelmed by available data. Another argument may be addressed for the use of such systems, and against sending the graduate student to the library for machine-readable items. That argument is that, when associating data, only about 30 or 40 items can be kept in mind at one time. It is quite true that one can start reading the Merck Index to sort out which chemical structures are associated with specified medical uses. One may do it for the first three pages, only to get swamped by the fourth page of an 1100-page volume.

The other source on which our work draws is the heuristic research methodology. Heuristic programming involves trial and error procedures rather than algorithms, and has become more practical with the advent of real-time interaction. The only condition for the practice of heuristic programming is that the practitioner himself sits down at the console and interacts with the computer. Only the practitioner can quickly recognize errors in his own specialty.

In the more usual deductive method of working, we formulate a specific hypothesis and then gather data which will validate or invalidate the hypothesis. By contrast, in the heuristic approach we begin with a comprehensive group of data such as a data base and we examine this data base repeatedly and interactively with new hypotheses. This is essentially an inductive technique. We have forgotten today, when the deductive technique is almost second nature among scientists, that some of the greatest advances even in the hard sciences were made by inductive techniques. I refer here to such seminal events as the formulation of the periodic table by Mendeleev and, going back even further, the derivation of a hierarchical organization of all animals and plants by Linnaeus in Sweden. A recent paper ($\underline{2}$) has listed other examples of important rules arrived at by the inductive process: Kepler's third law, Ohm's law, Prout's hypothesis, Balmer's formula, and others.

I might add that induction does not prescribe a priori the form in which data should be organized. Dr. Perone has been talking about this at this symposium ($\underline{3}$) and he makes a strong point for not prejudging the form which data should fit. Herman Chernoff, the statistician, observes that chemists may well be throwing away as much as 95% of the information contained in their data simply because they prescribe the form in which the data should be calculated ($\underline{4}$). For example, we express much of our spectroscopic data in the form of spectra and we expect to see certain peaks and valleys even if we plot them in a derivative manner. For some kind of data a spectral form may not be the form which gives the greatest amount of information inherent in those data and that is what we have to be prepared for in thinking about things in this way. Perhaps it is the freedom of form in the heuristic, inductive approach which is one reason why these methodologies may be less familiar to physical scientists than they are to behavioral scientists.

This heuristic approach has spawned a large number of clustering techniques in which Dr. Perone and some of his more chemometric colleagues have been active as far as chemistry is concerned. Again, there are statisticians such as Solomon ($\underline{5}$) and others who construct hierarchical taxonomies by these processes. The key to using these techniques is the computer and the techniques up to now have been largely limited to the use of numbers. What we have done is to try to extend this inductive approach to machine-readable alphanumeric chemical data.

Clustering

I now wish to consider the process of clustering in a little bit more detail. The data, whether they are literature data or experimental data, are contained in a hyperspace and, while there are probably a lot of better definitions of clustering, the way I visualize clustering when I wish to explain it to chemists is that

the data being considered form a hyperspace and one drops various
two-dimensional planes, one at a time, with recognizable axes
through the data and sees whether the data group on one or the other
of these planes. Furthermore, the game is to see whether the
two-dimensional planes indeed have recognizable axes, that is,
whether that particular plane on which data group corresponds to an
orthogonal relationship of variables which is recognizable to us.
Solomon, whom I quoted before, has told the story that he was asked
how many times does he run a clustering program for a client in his
consulting practice. How does he know when he has run the clustering
programs sufficiently often? His answer is an "ah-so point" at which
the client finally recognizes two orthogonal variables which were
common but not necessarily related in his experience and which the
computer-aided clustering had revealed to be related.

We generally refer to hyperspaces with numeric data points as
being parametric hyperspaces. In those parametric hyperspaces a
distance measure is relatively easy to construct and this distance
measure or metric then permits a measurement of closeness to be
assigned to any two elements in a cluster. That way we can define a
cluster of elements in hyperspace with great accuracy because we can
tell how close the points are. Alphanumerics, on the other hand,
form non-parametric hyperspaces. Here no numerical parameters are
associated with any element and here the prescription of distance or
closeness is very much more difficult. As a matter of fact, in
alphanumeric data the dimensionality of the hyperspace may not even
be completely defined. In such open-ended spaces clustering becomes
a rather ill-defined operation. It is the purpose of this paper to
indicate some heuristic approaches to clustering in such open-ended
spaces and to show the formatting for their use. The generalization
of these techniques to other types of data bases may be fruitful and
is something that I am very much looking forward to, particularly
now with the availability not only of word processing software but
also of the great new extracting and associating power which is
available to us from CAS ONLINE.

The Data Base

I have described the data base which we used in the development of
these techniques previously (6-8) and will not repeat the
description or derivation of the data base. Suffice it to say that
it was derived from the Eighth Edition of the Merck Index, and that
it lists all of the synonyms and all of the medical uses for all
compounds in the Eighth Edition of the Merck Index which show a
medical use, some 3400 compounds in all. We used conventional
sorting programs and text processing programs in interrogating this
data base. One of the things which we found out early on was that it
was not necessary to code the structure of chemical compounds into
machine-readable form. Rather, the use of the Geneva System name of
each chemical compound for alphanumeric data processing was the
substance of some of our early papers. I will now describe three
different alphanumeric clustering methods which we derived in the
course of our work.

Clustering Method I

The first method took advantage of the fact that from the very beginning we placed at least two properties on the same computer line, for example chemical structure, in the form of the Geneva System name, and medical use. We did not format the data base very rigorously at the beginning because we were not at all sure of the information that lurked in the data base and therefore did not want to restrict ourselves by adopting formatting that would produce only the information that we could predict or guess was available from the data base. I should add that this kind of humility in the face of data is one of the characteristics of the heuristic, inductive approach. When I say that we placed two properties on the same computer line and mention chemical structure as one of these properties, I mean that we used chemical structure as one of the hyperspace coordinates rather than as a word to be used mainly for bibliographic retrieval. Since we had two properties on the computer line we were able to run the two properties against each other (Table I).

Table I. Formatting for Alphanumeric Clustering Method I.

	FIELD 1	FIELD 2	FIELD 3
LINE 1	Item 1	Property A	Property B
LINE 2	Item 2	Property A	Property B
.	.	.	.
.	.	.	.
.	.	.	.

In other words, we could pull out all of the computer lines with property "A", could see what kinds of properties "B" were associated with property "A", and then run the reverse search for property "B" to see whether we had missed any previous "A's" or not. In this way we obtained the intersection of two hyperspace axes and therefore a primitive form of clustering in an alphanumeric system.

What did we do with this primitive form of clustering? Well, we studied compounds active in the autonomic nervous system. We did that because my collaborator was a psychologist. We were particularly interested in the question of physiological versus behavioral effects of the same kinds of compounds. In other words, if some related compounds had a physiological effect, could one read between the lines of the information and see that they also had a behavioral effect? On the other hand, when a number of related compounds had a behavioral effect, could one read between the lines and also see whether they had a physiological effect or not? If that had been the only purpose of our work we would have quit it after six months because we very quickly found out the answer to the question is obviously yes (<u>6</u>). Let me illustrate with some of the indole–nucleus compounds that we examined early on in our work (Table II).

Table II. Adrenergic Effects of Some Indole-Nucleus Compounds,
Classified by Alphanumeric Clustering Method I.

```
                    Physiologically recognizable
                       hemostatic
                       antihistaminic
                       serotonin antagonist
                    Behaviorally recognizable
                       hallucinogenic
                       diuretic, antihypertensive
                       analgesic, antipyretic
                       tranquilizer
                              .
                              .
                              .
```

Some of these have physiologically recognizable adrenergic
effects. Other indole compounds have behaviorally recognizable
adrenergic effects. In this list of behaviorally recognizable
adrenergic effects one already sees the mixture with physiologically
recognizable effects. For example, the same compound that is an
antihypertensive is also a diuretic, the same compound that is an
analgesic is also an antipyretic, etc.

We also looked into the method of action of mescaline ($\underline{7}$). We
found when we searched for phenethylamines that mescaline is always
associated with sympathomimetics and other substances which stimu-
late the sympathetic nervous system such as anorexigenics.

Clustering Method 2

A second, more sophisticated clustering technique for alphanumeric
information proceeds as follows: The data base can be imagined as a
collection of items which are described by a set of properties.
Again, our properties in the Merck Index data base are chemical
structures as names, and medical use. Each item in the data base is
assigned a vector whose column elements indicate whether the item
has a given set of properties or whether it does not. For example,
a 1 indicates the presence of the property and a zero indicates the
absence of the property (Table III).

Table III. Formatting for Alphanumeric Clustering Method II.

	Property A	Property B	Property C	. . .
Item 1	1	0	1	. . .
Item 2	0	1	1	. . .
Item 3	1	1	0	. . .
.	.	.	.	. . .
.	.	.	.	. . .
.	.	.	.	. . .

We can now select a subset of the properties, i. e., some
particular medical uses or some particular chemical structures and
search the data base for those items which have all those properties

by the way, is the same way in which CAS ONLINE works. The computer tries to find all the items with a 1 in the appropriate columns. The clustering aspect of this procedure is that the more properties any two items have in common, the closer they should lie in hyperspace.

Besides similarity, it is also possible to consider dissimilarity as a clustering technique. There may be pairs of properties which are exclusive or almost exclusive, that is, items with one property almost never have the other property. Such a finding would suggest that this exclusiveness expresses a relation between the properties. A relation of exclusion might thus indicate that properties are opposed in some way. Here we can use values of +1, −1 and 0 to show presence of the property, the opposite property or neither.

As an example I cite our early searches on the chemical structure properties indole and ethylamine and the medical use properties sympathomimetic and parasympathomimetic ($\underline{7}$, $\underline{8}$). These searches revealed two sets of intersecting hyperspace axes. One of these intersections contains compounds with ethylamine structures (Property A in Table IV) which have the medical use property sympathomimetic; the other cluster indicates organic ammonium ions

Table IV.   Clustering of Ethylamines and Organic Ammonium Ions by Alphanumeric Clustering Method II.

	1 Similarity	−1 Dissimilarity
Property A	51 items	−
Property B	27 items	9 items

which are parasympathomimetics. The 51 sympathomimetics included the catechole-type nerve impulse transmitters as well as numerous compounds which mimic their action in the sympathetic nervous system. The 27 parasympathomimetics found include the nerve impulse transmitter acetylcholine as well as compounds which mimic their action in the parasympathetic nervous system. Searches on the chemical property organic ammonium ion turned up compounds which were not parasympathomimetics (Dissimilarity in Table IV). These compounds contain bulky side groups and function as skeletal muscle relaxants by blocking acetylcholine. Nine compounds of this type known by the medical use property curarimimetic were found in the search. The medical use term or property parasympatholytic, denoting inhibition of nerve impulse transmission in the parasympathetic nervous system is not used in the Eighth Edition of the Merck Index.

Dissimilarity (opposite effect) due to bulky side groups hindering the action of compounds containing the ethylamine moiety were also found. Again the data base lacked a descriptor indicating the medical use property sympatholytic. Examples of such compounds would be tranquilizers such as the reserpine alkaloids. An effective search strategy for compounds of this type has not been developed and consequently the space for them in Table IV cannot yet be filled in.

Clustering Method 3

A third and more quantitative method of clustering allowed us to
make four different cuts through the hyperspace of medical uses and
their frequency in the Merck data base. By cut I mean a projection
of the hyperspace onto one or more recognizable orthogonal axes.
This is equivalent to the process I was talking about earlier in
which we pass a plane with hopefully recognizable orthogonal axes
through the hyperspace. Here we are talking about two different sets
of properties then we did in the previous methods, in which we
talked about items having the properties chemical structure and
medical use. Here we talk about items having the properties medical
use and the frequency of that medical use in the data base. Two
parameters were used in making these cuts. First of all we counted
the number of medical use per item (i. e., per compound that is the
connection to chemical structure which may be exploited later). The
way in which uses/compound were tabulated is shown in Table V: a
more complete tabulation can be found in our earlier papers (9).

Table V. Formatting for Alphanumeric Clustering Method 111.
Derivation of Distribution Curve.

Uses/Compound					Freq.	Use Name
1	2	3	4	5		
201	24	1			226	antimicrobial
.	.	.	.	.	.	.
.	.	.	.	.	.	.
.	.	.	.	.	.	.

      If we sum the uses/compound, we get a frequency for each
medical use. If we plot the number of different uses against the
frequency of that use on probability paper or log-log plot, we get a
distribution curve. The distribution curves for four different
subsets of the data in appear in (9). Each shows that the data form
a fairly straight line over two orders of magnitude. The curves
represent an attempt to fit the points with a normal, rather than a
log-normal, distribution. It is obvious that the curves do not fit
the points and that therefore the medical use distribution function
is log-normal rather than normal. My understanding is that this
log-normal distribution is typical of natural text data bases
despite the highly specialized character of the medical use data
base.
      When we take away the first column of Table V, we construct a
table which shows various pairs of uses as a function of how often
these use pairs occurred on a uses per compound axis (Table VI).

Table VI. Formatting for Alphanumeric Clustering Method III.
Derivation of Use Pair Taxonomy.

Uses/Compound				
2	3	4	5	
29	12	1	14	pair 1
27	5	1	0	pair 2
.	.	.	.	.
.	.	.	.	.
.	.	.	.	.

For example, the second use pair occurs 27 times as a pair but also in multiple combinations, five times with a third use and once with two other uses. A more complete version can be found in our earlier papers ($\underline{9}$). The use of pairs in that table can be grouped by closeness of relation. For example, the first nine use-pairs are all either analgesics or sedatives, the next four are all either antiseptics or astringents. In a taxonomic sense these grouped use pairs represent a cluster; the seven we found are listed in Table VII.

Table VII. Use Pair Clusters in Merck Index Date Base, Derived by Alphanumeric Clustering Method III.

Analgesic-Sedative
Antiseptic-Astringent
Cardiotonic
Diuretic-Antihypertensive
Adrenocortical steroid
Parasympathomimetic
Sympathomimetic

This kind of clustering, a second cut through the hyperspace, can lead to extrapolation, dividing additional uses for existing compounds, and to structure-activity relationships.

The mean uses per compound in Table V is 1.7. The mean uses per compound can be counted for each of the uses. It was shown ($\underline{9}$) that mean uses per compound is distributed statistically and the outliers on each side could be identified. The ones with low mean uses per compound were highly specific uses such as antimicrobial, antineoplastics, antihistaminics, estrogenics, etc. The ones with high mean uses were the ones which were present in dominant use pairs or clusters such as diuretics, vasodilator, CNS depressant, etc., or which had old imprecise use descriptors such as sudorific, diaphoretic, dermatoses, etc.

Certain medical use qualifiers were coded into the data base. These medical use qualifiers form a fourth cut through the hyperspace, which again offers a statistically significant way of distinguishing between kinds of medical uses. As was seen from the distribution function for uses/compound, the X qualifier (null character-no additional information) showed a mean uses/compound which is not significantly different from the entire data base. This was also true of the Z qualifier (additional information). However,

the H (has been used) and F (formerly used) qualifiers show
significantly more mean uses per compound, and the set of
experimental qualifiers show significantly less mean uses per
compound.

Conclusion

I have reviewed three methods of clustering text data. Obviously the
field is in its infancy, and there is much work that can be done.
However, the power of this approach is demonstrated by its
simplicity and usefulness. I believe that the world of physical data
has exploded so much in the last 20 years that the inductive
approach can be helpful in organizing it for extrapolation,
interpolation, and planning, as well as for recognizing interactions. In that process the heuristic use of computers can be of
tremendous help if we only permit it to help.

Literature Cited
1. Dessy, R. "The Rational Electronic Laboratory," presented at
    the 186th ACS National Meeting, Washington, D. C., September 1984.
2. Bradshaw, G. F.; Langley, P. W.; Simon, H. A. Science 1983,
    222, 971-5.
3. Perone, S. Chapter 9 in this book.
4. Chernoff, H., personal communication.
5. Solomon, H. "Numerical Taxonomy"; Technical Report No. 167,
    Stanford University Department of Statistics, Stanford, CA,
    Dec. 1970.
6. Gloye, E. E.; Marcus, R. J. Science 1970, 169, 88-91.
7. Marcus, R. J.; Gloye, E. E. J. Chem. Documentation
    11, 163 (1971).
8. Marcus, R. J.; Gloye, E. E.; Florance, E. T. Computers &
    Chemistry 1977, 1, 235-241.
9. Marcus, R. J.; Florance, E. T.; Gloye, E. E. In "Retrieval of
    Medicinal Chemical Information"; Howe, W. J.; Milne, M. M.;
    Pennell, A. F., Eds.; ACS SYMPOSIUM SERIES No. 84, American
    Chemical Society: Washington, D. C., 1978; pp. 39-57.

RECEIVED June 1, 1984

Coping with the Information Explosion Provided by Modern Chemical Instrumentation

SAM P. PERONE

Chemistry and Materials Science Department, Lawrence Livermore National Laboratory, Livermore, CA 94550

Modern chemical instrumentation is capable of generating enormous amounts of data in very short periods of time. It is clear that a major task of scientists for the near future is to develop techniques to utilize more effectively this capability, in order to avoid the typical dilemma of being buried in data with little or no perspective of the information content. Thus, there are three key developments that must be pursued: definition of "information content"; identification of methods to correlate instrumental parameters with information content; and development of tools for the instrumental enhancement of information content and the efficient extraction of information from data. These developments should allow the evolution of "smart instruments", perhaps guided by artificial intelligence principles. This paper will describe some of the principles and tools that have already been developed, and will identify the areas where work needs to be done.

Modern instrumentation for chemical analysis, because of the incorporation of digital computer systems, allows the generation and collection of immense amounts of data. This is facilitated by computer control of experimental variables and high-speed collection of multiple channels of data. This in turn allows complex measurement principles to be implemented, with correspondingly complicated multivariate analysis.

Unfortunately, the data explosion that has accompanied the evolution of modern chemical instrumentation has not provided a corresponding information explosion. This is because relatively little attention has been paid to the development of techniques for optimization of information content, or for enhancement and extraction of information. It is not uncommon to observe a scientist buried in a data printout from an experiment, manually scanning columns of data, calculator in hand, attempting to extract useful information.

0097–6156/84/0265–0099$06.00/0

It is time to turn our attention to developing more effective methods for obtaining information from complex experimental systems. The first step involves the definition of generic concepts of information content which are independent of the specific instrumental system. This is a task which has been surprisingly neglected in the past. The very simplest concepts which must be defined include:

o   informational goals
o   information content
o   information enhancement

The next step is to apply the basic principles of information theory, signal processing theory, multivariate data interpretation, and adaptive instrumental control in order to enhance and effectively extract information.

Information Goals

The primary requirement in the process of information enhancement is to define the informational goal(s) associated with a set of experimental measurements. Equally important is the definition of an appropriate measure of the degree to which the informational goal is achieved. Some generic qualitative informational goals and their respective figures of merit might be:

GOAL	FIGURES OF MERIT
concentration	accuracy/precision
resolution	peak separation/peak width
sensitivity	detection limit/response slope
matrix effects	linearity/interference effects

In addition, it is possible to define qualitative informational goals. These might include:

o   identification of chemical components
o   classification of materials/properties
o   establishment of chemical mechanism.

Corresponding figures of merit for the qualitative informational goals can be defined in terms of statistical accuracy by evaluation with systems of known properties.

Information Content

This concept is one of the most difficult to quantitate. There are some relatively explicit definitions of information content for electronic communications. (For example, the Nyquist theorem defines the minimum sampling rate required in order to preserve the maximum frequency information in a periodic signal. And, the relationships between digital encoding formats and information content of a data base can be quantitated.) However, for the general problem of evaluating the results of instrumental measurements of chemical systems, the definitions for information content of data are very clear.

One goal of our research program is to develop explicit and quantitative definitions of information content which may be useful for chemical instrumentation systems. These will be based on the principles of information theory, sampling theory, and signal processing theory. At this time, however, we can describe an

empirical approach to evaluation of information content which we
have found very useful.

This approach involves the following steps:

- o define the "desired information" (informational goal(s))
- o define a figure of merit for goal achievement (e.g.,
     accuracy, precision, reliability, etc.)
- o empirically determine "information content" from the
     relationship:

$$[\text{INFO. GOAL}] = f[\text{INFO. CONTENT}] \tag{1}$$

From the above statement the information content of a chemical
measurement system can be evaluated by studying the effects of
experimental factors on the degree of achievement of the
informational goal(s). This is elaborated below.

Information Enhancement

An empirical procedure can be defined for the enhancement of
information content. First, it must be recognized that the
achievement of desired informational goal(s) depends not only on
the inherent information content of data, but also on the data
management and analysis procedures. This is expressed in
Equation (2):

$$[\text{INFO. GOAL}] = f[\text{CONTENT, MGMT, ANALYSIS}] \tag{2}$$

Thus, to examine the relationship between information content
and experimental factors, it is necessary to maintain consistent
data management and analysis procedures. Then, one can assume a
direct relationship between the achievement of informational goals
and information content as implied in Equation (1).

A study designed to determine the effects of experimental
factors on information content might be based on the relationship
defined by Equation (3):

$$[\text{INFO. CONTENT}] = f[\text{MEASUREMENT PRINCIPLES,}$$
$$\text{EXPTL DESIGN,}$$
$$\text{EXPTL PARAMETERS}] \tag{3}$$

Procedurally, one could vary any of the experimental factors in
Equation (3) and evaluate the effects on information content under
conditions where Equation (1) applies.

In order to clarify the general concepts defined in the above
sections, the following sections will describe an experimental
study which followed those principles in order to achieve specified
informational goals.

Electrochemical Structural and Activity Classifications

The classification of chemical structure using electrochemical
techniques, is a challenging problem. Voltammetric responses lack
fine structure and probably will never compete with spectroscopic
methods in qualitative analysis. The complex dependence of an
electrochemical response on many variables, and theoretical

problems in relating structure to electrochemical activity, make qualitative voltammetric analysis even more formidable.

Even though the difficulties in qualitative electroanalysis are great, the rewards of developing a reliable means of structural identification through electroanalysis would also be great. Due to recently developed miniaturization techniques, electrodes are the most promising probes of _in vivo_ chemical species. Carbon fiber electrodes may be implanted within a single cell or neuron (_1_). Electrochemical detectors in liquid chromatography are becoming very important because of their high sensitivity and selectivity. Quantities of electroactive material in the picogram range have been analyzed. Osteryoung, et al. (_2_) have demonstrated the feasibility of scanning the potential of a liquid chromatographic electrochemical detector, so the development of qualitative voltammetric methods would open up the possibility of the characterization of eluants that are 1000 times less concentrated than those which can be analyzed by spectroscopic techniques.

Linear-free-energy relationships have generally been the most useful expressions for relating structure to electrochemical activity in the past. A substituent group will have a characteristic effect on the free energy of an electrochemical reaction occurring in its vicinity. This effect may occur through electron withdrawal, electron donation, or it may be steric in nature. In any case, the effect may be quantified through the use of Hammett substituent constants. For a given class of electrochemical reactions, there will be a linear relationship between $E_{1/2}$ and the substituent constants σ (_3_).

There are two main problems in the use of linear-free-energy relationships. The first and largest problem is the determination of the reaction series to which an unknown belongs. Such a deduction from electrochemical behavior is not straightforward. Furthermore, there may be several reaction series which may be constructed for a class of compounds depending on solution conditions. The slope of the $E_{1/2}$ vs σ plot would be different at high pH's due to a change in the mechanism of reduction.

The second main problem is that there is often not enough $E_{1/2}$ separation for different substituents or substituent combinations to allow for confidence in identification, especially when experimental reproducibility is low due to uncontrolled matrix effects. The consideration of more information than $E_{1/2}$ would clearly be helpful.

Because pattern recognition is well suited to the consideration of large amounts of information and to making use of obscure relations, we have applied it to chemical structure identification from electrochemical data. The main questions have been what data should be collected and how much?

Burgard and Perone (_4_), used staircase voltammetry to analyze 29 compounds belonging to four different electroactive group/skeleton combinations. The classes examined were aromatic-nitro, aliphatic-nitro, aromatic-aldehyde and aromatic-aliphatic-ketone. Fortuitously these classes were almost completely separated on the basis of peak potential; but this feature alone cannot be considered sufficient for many identification problems. Thus, the voltammograms were examined for any shape information which might characterize a particular

electroactive group or the skeleton to which it was attached. It
was found that the change in peak shape with scan rate produced
fair classifications (70% correct), but that complete separation of
the classes was not possible for the experimental conditions and
compounds which were chosen. The results suggested that the
information content of the electrochemical data base should be
increased for more reliable structural classifications.

The work described below by Byers, Freiser, and Perone (5,6)
represents an attempt to define quantitatively the information
content of electroanalytical voltammetric data with regard to
structural and activity classifications. The general principles
defined in the introductory sections of this paper were followed.

Results and Discussion

Ichise, Yamagishi and Kojima (7-9) have proposed the simultaneous
determination of complete E-i-c and C_{dl}-E-c patterns (c = surface
concentration) and have published several papers on instrumentation
and data compression algorithms for reaching that goal. E-i-c
patterns were generated by applying a pseudo-random waveform to the
cell and monitoring the current response. The surface
concentration of the depolarizer was calculated from the current in
an analog fashion with an "$s^{-1/2}$ module" which eliminated the
effect of diffusion. C_{dl} was obtained by applying a high
frequency 10 mV sinusoidal wave to the cell and measuring the
amplitude of the 90 degrees out-of-phase component of the current.

The idea of obtaining double-layer capacity information may be
fruitful. The capacitance of the double layer is dependent on
adsorption of the analyte, and the strength and potential
dependence of adsorption may indicate the presence of certain
functional groups (10). π-electron interaction between adsorbed
molecules and the electrode surface has a characteristic influence
on the adsorption behavior of organic substances (10), and specific
interactions between the analyte and some other molecule or ion
within the double layer may also be helpful in identification
(11,12). Some adsorbed organics will inhibit the reduction of
metal ions, while others, through the so called "cap-pair" effect
will accelerate reductions (13).

The use of a potential-step technique such as cyclic staircase
voltammetry represents a simple alternative to Ichise's method (8)
of obtaining information on both adsorption and electron transfer
kinetics. The current decay immediately after a step is primarily
capacitive while current at later times is almost totally due to
electron transfer reactions. Thus, by measuring the current at
several times during each step and by changing the scan rate,
information on both the kinetics of the electrode process and the
differential capacity can be obtained with a single sweep.

As is true with cyclic linear sweep voltammetry, the reversal
of the scan is important in detecting chemical reactions which
succeed the electron transfer step. Immediate repetition of a
cyclic scan may detect products which have been generated in the
reverse scan of the first cycle.

One additional parameter which can be explored is the "drop
hang time". This refers to the time period between the creation of
a stationary mercury drop and the beginning of the first staircase

scan. During the waiting time, a potential can be applied. This
variable was investigated in our work to see if there was any class
specific information in the kinetics of adsorption.

Another source of structural information is the electrochemical
response of the analyte to chemical perturbations. Changes in
solution conditions have been useful in classical studies of
structure-activity relationships. Exploration of a variety of
solutions will help define the best conditions for particular
classification problems.

All of the experimental and solution variables which have been
examined systematically in our classification studies are listed in
Table I. The determination of the effect of each of the seven

TABLE I.   Variable Levels for Factorial Design to Study Structural
Effects on Voltammetric Data

VARIABLE NUMBER	VARIABLE	LOW LEVEL (−)	HIGH LEVEL (+)
X_1	% Ethanol	0.5 %	9.5 %
X_2	pH	8.0	5.1
X_3	Surfactant Concentration	0	1.4×10^{-5} M
X_4	Number of Cycles	1	2
X_5	Scan Rate	0.25 V/s.	1.0 V/s.
X_6	Drop Hang Time	0.2 s.	30 s.
X_7	Sampling Time	30% of step $(\alpha'=.7)$	end of step $(\alpha'=.007)$

variables is difficult without good experimental design. To
characterize all main effects and all interactions one could
arrange the experiments by a factorial design (14). For the seven
variables considered here, 128 runs would be needed for each
compound. The large number of runs can be avoided by using a
saturated fractional factorial design (15) in which the main effect
of all seven variables can be investigated in only eight
experiments. By running a second fraction, in which all variable
levels have been reversed from their state in the first fraction,
all confounding between the main effect of variables and the
interaction of two variables will be eliminated. Higher order
interactions (the interaction of three or more variables) may still
be confounded with the main effects, but in most cases such
interactions are relatively small in magnitude.

In our work (5,6), a fractional factorial design was used as
described above. In addition, one of the experiments run early in
the analysis of each compound is repeated near the end of the
analysis to determine instrumental precision and to detect any
decomposition of the sample. This makes a total of 17

voltammograms which must be taken for each compound. These
experiments yield 17 current-voltage and 17 differential capacity
curves for each compound.

Graphical analysis of the error involved in the calculation of
variable effects was done for several nitroaromatics and
nitrodiphenyl ethers (5). It was discovered that all of the
variables chosen for study had significant effects on the Faradaic
responses of the compounds examined. The magnitudes of the effects
and the shapes of the effect curves were quite different,
indicating that redundant information was not recorded. All of the
variables also had a significant effect on the differential
capacity curves of strongly adsorbed species, but some of the
effects could not be distinguished from noise for more weakly
adsorbed compounds. Only pH, number of cycles and % ethanol had a
significant effect on the capacitance response of both weakly and
strongly adsorbed organics.

Since the variables chosen and the levels over which they were
changed seemed to be appropriate for most compounds from a
signal-to-noise perspective, the variable effects were further
examined for any information which might be useful in structural
classifications. Forty-five compounds representing three major
structural classes were chosen, and features derived from the
variable effects were tested for predictive ability (6). Class 1
consisted of 19 nitroaromatics containing a single benzene ring;
Class 2 contained nine nitrodiphenylethers, and Class 3 consisted
of 17 azo compounds. The classes were completely overlapped in
potential, and all compounds were reduced by the same number of
electrons, so the identification of the classes from their
voltammetric behavior was not a trivial problem.

In terms of the concepts defined in the introductory sections,
the informational goal of this study was "structural
classification". The figure of merit for achievement of this goal
was "classification accuracy" for examination of a data base
containing a large number of items of known class. The
experimental parameters were varied systematically according to a
fractional factorial design. Ultimately, it was desired to
establish what combination(s) of experimental parameters produced
electroanalytical data with the highest information content,using
the figure of merit defined above.

The pattern recognition analysis revealed that all of the
variables produced structural-specific information. Most of the
information was found in the Faradaic responses. Changes in the
Faradaic responses with the number of cycles gave the highest
classification accuracy of 93.3%. Scan rate changes yielded 89%,
while pH, surfactant and drop hang time all produced classification
accuracies of 84%. Changes in Faradaic response with % ethanol and
sampling time appeared to contain the least structural information,
giving classification accuracies of 66.7 and 75.6%, respectively.
As was expected from the signal-to-noise analysis, the effects of
the several variables on the capacitive responses were much poorer
structural predictors. Classification accuracies ranged between
60.0 and 75.6%.

Although changes in differential capacity responses caused by
changes in the experimental variables were not very helpful, the
shapes of differential capacity curves which were obtained under

the same experimental conditions were excellent structural
descriptors. Using shape features derived from differential
capacity curves taken under one set of experimental conditions,
93.3% classification accuracy was achieved. Four other sets of
experimental conditions yielded over 90% classification accuracy.

An interesting sidelight of the organic structural
classification study was that herbicidal activity could also be
predicted (6). The nitrodiphenylethers could be divided into
compounds which were strong herbicides and those compounds which
showed little or no herbicidal activity. Both Faradaic and
capacitive responses could be used to separate these classes for
over half the experimental conditions examined. As was found in
the classification of structure, capacitive factorial features
performed somewhat better than Faradaic factorial features. It
also appeared that classifications of herbicidal activity using
Faradaic factorial features could be improved considerably by
working at high pH and without surfactant present. The information
content of Faradaic or capacitive variable effects data could be
improved by variations in % ethanol.

The ability of voltammetric responses to predict the herbicidal
activity can be explained by the mechanism of herbicidal action for
the nitrodiphenylethers. It is thought that these compounds are
involved in the initiation of destructive free radical reactions
with the phospholipid molecules which make up cellular membranes
(16). Since the first step in the reduction of aromatics at the
mercury electrode also involves the formation of radical species
(17), some correlation between herbicidal activity and voltammetric
behavior is not surprising.

Conclusions

The experimental study described here illustrates how the
application of the principles of information enhancement can
significantly improve chemical analysis. In this case we have
established the optimum conditions for obtaining structural or
activity information from voltammetric electroanalytical data.
Moreover, it is clear that the informational goal(s) will dictate
the most favorable choice of experimental conditions. It is also
interesting to observe that the most useful experimental conditions
--- such as the enhancement of surface interactions --- are not
necessarily those which are traditionally valued most highly in
voltammetric studies. This result points up another valuable
benefit of an objective systematic information enhancement study.
Finally, it should be observed that the principles and general
methodology described in this work are generic and should be
applicable to any chemical instrumental systems.

This work supported by the Office of Naval Research and the
U.S. Department of Energy Contract W-7405-ENG-48 Lawrence Livermore
National Laboratory.

References

1. Ponchon, J. L.; Cespuglio, R.; Gonon, F.; Jouvet M.; Pugol,
    J. F., Anal. Chem., 51, 1483 (1979).

2.   Samuelsson, R.; O'Dea J.; Osteryoung, J., Anal. Chem., 52, 2215 (1980).
3.   Zuman, P., "The Elucidation of Organic Electrode Processes"; Academic Press: New York, 1969, Chapter 2.
4.   Burgard D.; Perone, S. P., Anal. Chem., 50, 1366 (1978).
5.   Byers, W. A.; Perone, S. P., Anal. Chem., 55, 615 (1983).
6.   Byers, W. A.; Freiser, B. S.; Perone, S. P., Anal. Chem., 55, 620 (1983).
7.   Ichise, M.; Yamagishi H.; Kojima, T., J. Electroanal. Chem, 94, 187 (1978).
8.   Ichise, M.; Yamagishi, H.; Oishi, H.; Kojima, T., J. Electroanal. Chem., 106, 35 (1980).
9.   Ichise, M.; Yamagishi, H.; Oishi, H.; Kojima, T., J. Electroanal. Chem., 108, 213 (1980).
10. Damaskin, B. B.; Petrii, O. A.; Balrakov, V. V., "Adsorption of Organic Compounds on Electrodes"; Plenum Press: New York, 39-40 (1971).
11. Gupta S.; Sharma, S., Electrochim. Acta, 10, 151 (1965).
12. Dutkiewicz, E.; Puacz, A., J. Electroanal. Chem., 100, 947 (1979).
13. Sykut, K.; Dalmata, G.; Nowicka, B.; Saba, J., J. Electroanal. Chem., 90, 299 (1978).
14. Hendrix, C. D., CHEMTEC, 9, 167 (1979).
15. Box, G. E. P.; Hunter, W. G.; Hunter, J. S., "Statistics for Experimenters"; John Wiley and Sons: New York, 1978; Chapter 12.
16. Orr, G. Ph.D. Thesis, Purdue University, W. Lafayette, IN (1981).
17. Kastening, B.; Holleck, L., J. Electroanal. Chem., 27, 355 (1970).

RECEIVED June 20, 1984

The Universe Is Stochastic and Nonlinear

LARRY M. STURDIVAN and BARBARA A. B. SEIDERS

Chemical Research and Development Center, Aberdeen Proving Ground, MD 21010

It is perhaps in the nature of man to look for determinism in the
universe. In a primitive society, necessarily very close to
nature, sound and movement in inanimate objects were associated with
unseen, and potentially dangerous, animal life: a wolf brushing past
the undergrowth, or a snake slithering between the rocks. It was
only natural that man would feel compelled to invent invisible
spirits to move the wind and water. Today while we know what moves
the wind and water, their apparent randomness continues to trouble
us. We build in the paths of flood and hurricane, and continue to
blame the damage they cause on their unpredictable nature. The
implication is that if we knew <u>all</u> the factors involved, we could
predict the weather long enough in advance to do something about it.

After millennia of debating whether nature is in its essence
deterministic, and therefore predictable, the answer is still not
known. Even Albert Einstein who pioneered work in statistical
mechanics was quoted <u>(1)</u> as saying: "Quantum mechanics is very
impressive. But an inner voice tells me that this is not the real
Jacob. The theory has much to offer, but it does not bring us
closer to the secret of the Old One. At least I am convinced that
He does not throw dice."

As long as our knowledge of the nature of the universe is
obtained by observation rather than by inspiration, we would argue
that the question cannot be answered. Limits on our ability to
observe details at the atomic and subatomic levels, as expressed by
Heisenberg's Uncertainty Principle, put the answer permanently
beyond our grasp.

However, in the context of the everyday laboratory the question
is moot. It may be that the immutable laws of a deterministic
universe dictated that in the middle of a star at the edge of the
universe millions of years ago an atom was stripped of its electrons
and sent our way at just less than the speed of light. But when
that "cosmic ray" crashes through out cloud chamber, ruining our
experiment, we have no choice but to regard it as a chance event.
Even if we had an infinite capacity to store facts about the present state of the universe and had them all in place (universal
data base) and if we had an infinite processing rate (the ultimate
computer), we still would need an exact model of the universe

(method of combining the facts) to predict the future unerringly.
However, it is likely that the only exact model of the universe is
the universe itself. It is inevitable that any model we construct
of some piece of the universe will have artifacts that have no
analog in reality. Since we cannot find exact models, the best we
can do is to build models which are useful in the context within
which we wish to employ them. It may be that the concept of
probability is itself an artifact which man has invented to express
the uncertainty which arises from excluding relevant factors from
the model which are infeasible or impossible to measure or whose
influence is unsuspected. If so, it is an artifact which is often
very useful when employed properly.

Another artifact whose appealing simplicity has resulted in its
overuse is the concept of linearity. In fact, the term is used to
express several closely related concepts. Originally, the term
linear meant a straight line. Mathematically, the equation for a
straight line is the same as for a strictly proportional relationship:

$$y = a + bx$$

This was extended to include additive proportionality models:

$$y = a + bx + cz$$

or, most generally,

$$y = \sum_i b_i x_i$$

where the b_i are (unknown) proportionality constants and the x_i are
independent variables or functions of independent variables and
known constants. The latter equation includes polynomials as well
as functions of several variables. It is termed linear with respect
to the unknown b_i's. In space, such functions are no longer equivalent to straight lines but to planes and hyperplanes in n-
dimensional space. When we say the universe is nonlinear we mean
that such equations are seldom very useful models for extrapolating
or interpolating natural systems. In a very real sense, the phrase
could also be applied to space itself. In relativistic terms,
because of the curvature of space-time, the geodesics that light
follows are not straight lines. Thus, both literally and figura-
tively, the universe is nonlinear. If one is dealing with a small
enough piece of it, a linear approximation might be useful. If one
is lucky, the variance might be small enough that it could be considered deterministic. If one is _very_ lucky, it might be adequate
to model it as both linear and deterministic.

But what do we do when we aren't so lucky? Most of the models
that physical scientists have proposed acknowledge the nonlinear
nature of nature, but consider it deterministic. Statisticians
acknowledge the stochastic nature of the universe, but have only
taken a few tentative steps beyond the "general linear hypotheses".
Not much has been done in that difficult area where a clearly non-
linear phenomenon involves a significant element of chance.

However, avoiding the problem is like the drunk who looks for his keys under the streetlight because the light is better there than in the middle of the block where he lost them.

There are some techniques that are clearly useful and principles which can be generally applied to such problems. These can probably best be introduced by means of some examples. The first is from our experience in personal protection. The penetration of fabric body armor by a ballistic projectile is clearly a stochastic phenomenon. For any particular vest-projectile combination, there is a span of velocity in which penetrations and nonpenetrations are mixed. The probability of penetration within this zone is influenced by at least the following factors:

V - velocity of the projectile
M - mass of the projectile
A - size of the projectile
t - thickness of vest
T - tensile strength of vest material

Because the vest is made of several layers of cloth, the measurement of thickness is a difficult task. (How much air do you attempt to squeeze out from between and within the layers?) The most consistent method is to calculate the thickness equivalent to that which the vest would have if it were a single solid layer, i.e.,

$$t = \frac{\text{mass per unit area of cloth}}{\text{density of the material}}$$

The size of the projectile is well represented by its mean presented area, i.e., the mean area of its shadow cast on a plane averaged over all possible orientations. An appropriate scaling model can be derived for penetration, although the derivation is beyond the scope of this paper. The result may be expressed as

$$x = \frac{\frac{1}{2} mv^2/At}{T}$$

where the ratio X is dimensionless. According to the principle of similitude (2) scaling laws must consist of combinations of dimensionless ratios. Not all dimensionless ratios constitute legitimate scaling laws, but all legitimate scaling laws can be expressed as a combination of dimensionless ratios. It is postulated that equal values of the variable x would give rise to equal probability of vest penetration. Figure 1 shows a plot of penetration data from a number of projectiles fired against vests of various thicknesses. The data points represent mean penetration velocities, derived from a number of inpacts, for each vest/projectile combination. It is plotted on logarithmic axes to equalize variance. Energy ($\frac{1}{2}MV^2$) is plotted against the other variables to show the spread of data in energy. Because logs are plotted against logs a line of slope 1 represents a contour of equal probability (equal values of x). The tensile strength T is missing from these axes because the data were scaled for a constant T before plotting (i.e., it is included implicitly). After an appropriate scaling model is found, a probability function may be fitted to the data. The function fitted should be appropriate for the situation. If there is no theoretical basis for choosing one over another, then the choice can be made on convenience. For

                                              COMPUTERS IN THE LABORATORY

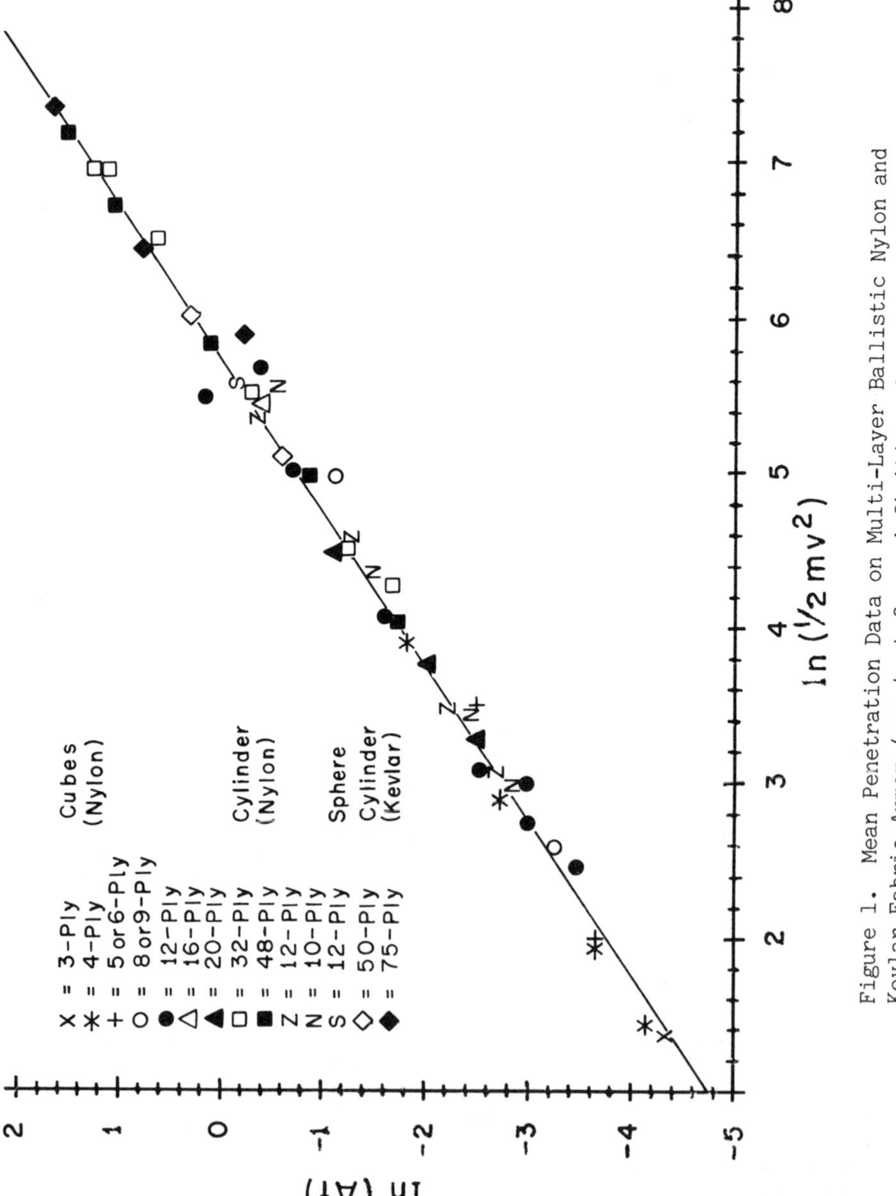

Figure 1. Mean Penetration Data on Multi-Layer Ballistic Nylon and Kevlar Fabric Armor (see text for a definition of variables).

fitting dichotomous data to the Logistic function, a mathematically tractible distribution, the method of Walker and Duncan
(3) is convenient. Figure 2 shows the typical S-shaped probability
distribution resulting from fitting the Logistic function:

$$p = \frac{1}{1+e^{-(a+b\ \ln\ x)}}$$

to the dichotomous data on penetration. The straight line of
slope 1 in figure 1 is actually the 50% probability contour from
the equation fitted to raw data. It is not a least squares fit to
the means plotted on the figure.

The second example is from a mixed biological/physical problem.
It deals with the probability that blunt trauma to the chest or
abdomen would be lethal to man. It has been used to assess the
hazard of large ballistic projectiles moving at moderate velocity,
the hazard behind body armor which has stopped a handgun bullet, etc.

The scaling model, which again is too lengthy to derive, is
(4)

$$x = \frac{\tfrac{1}{2}\ MV^2}{W^{1/3}td}$$

where   M - mass of the projectile

       V = velocity of the projectile

       W = mass of the individual

       t = thickness of the body wall over the vulnerable organ

       d = $\sqrt{A/4}$ = the effective diameter of the projectile

       A = mean presented area

Notice that if the constants

    ρ = mean density of the individual

    T = tensile strength of the tissue

were included, the product would be a dimensionless ratio comparable
to that of the previous example; i.e.,

$$x = \frac{\tfrac{1}{2}\ MV^2}{(\tfrac{W}{\rho})^{1/3}td\ T}$$

As in the previous model, the factors assumed to remain constant,
ρ and T, are assumed to be absorbed in the curve fitting constants
when fitted to the probability function. Figure 3 shows how well
the model fits the mean data. A plot of the probability curve
would be exactly like Figure 2 with a change in scale.

Given these introductory examples of applied stochastic models,

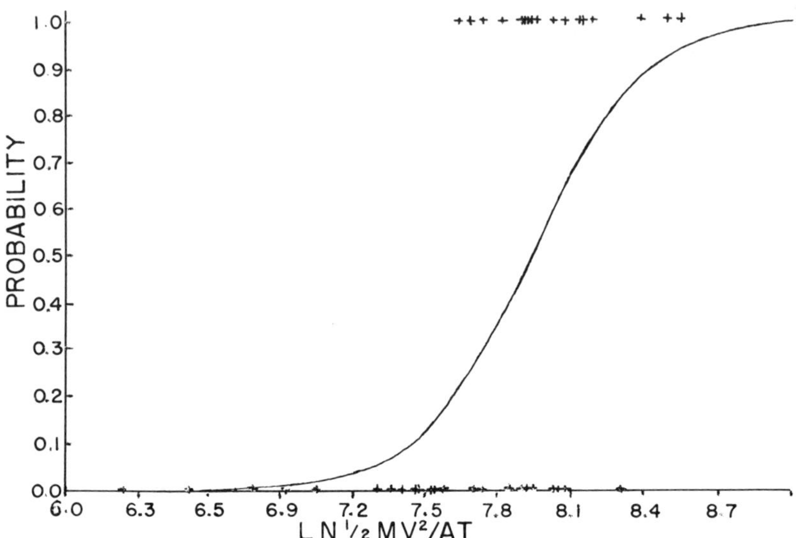

Figure 2.   The Probability of Penetrating Fabric Armor as a Function of the Model Variable, x.

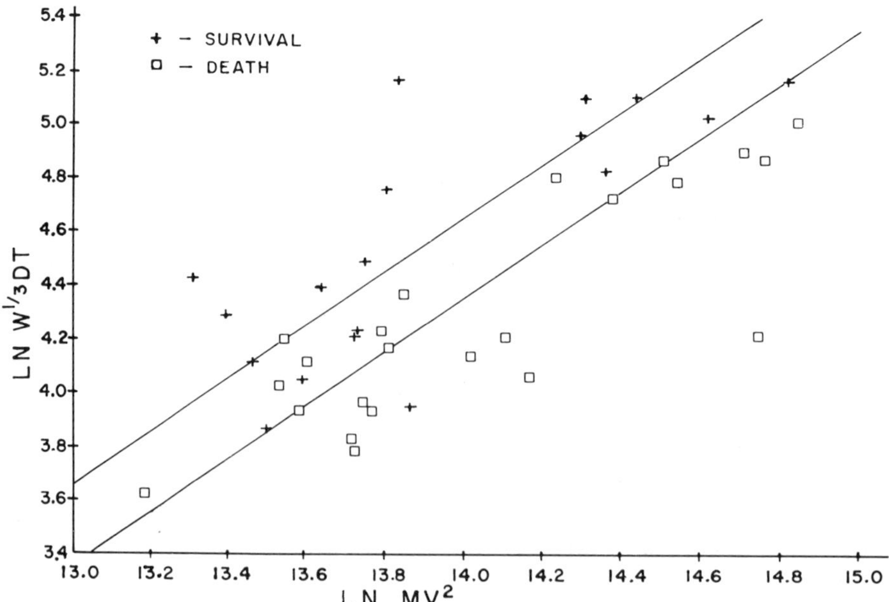

Figure 3.   Vulnerability of the Thorax to Blunt Trauma (see text for a definition of variables).

we can discuss in more detail some of the techniques and
principles which are particularly useful in deriving and fitting
this type model.

One of the most useful modeling andscaling techniques to be
found is Dimensional Analysis, embodying the principle of similitude (5,6) so-named by Galileo in the 17th century and given a
formal framework in 1822 by Fourier. The rules for manipulating
the fundamental units of measure which Fourier proposed has evolved
into the modern technique of dimensional analysis. The major addition in modern times is the Buckingham Pi Theorem by means of which
dimensionless ratios of the type used above may be derived. It
should be noted that in each of the examples the model shown was not
the first tried. For dimensional analysis to produce useful results
the whole set of relevant variable must be included, the proper
dimensionless ratios must be found, and, finally, the best method of
employing those dimensionless ratios in a model must be determined.
Dimensional analysis is just one method of normalizing the data;
i.e., making it independent of the units of measure. It is, however,
the best. Another method which is widely employed is to subtract
a known or inferred population mean from the individual datum and
to divide by the population standard deviation.

Once a scaling model has been found the scaled data should be
examined carefully to ascertain that the variance is equal over the
domain of the data. If not then a suitable transform must be found
to equalize the variance. Otherwise, no single stochastic model
will accurately reflect the probability of an occurrence of the
"event" in question over the data domain, much less for an extra-
polated prediction. For example, if the standard deviation is
proportional to the mean, a very common situation in nature, the
variance is equalized by taking the log of the model variable. This
is the case for both of the above examples, where the probability
model was fitting to ln x rather than x itself. Suitable trans-
formations for other common situations, as well as a general method
for finding transforms is given by Johnson & Leone (7).

When a suitable scaling model has been found and equal
variance confirmed or obtained, a probability function is fitted
to the data. For dichotomous data, the Gaussian (probit) or
Logistic (logit) functions are the most common mathematical functions used. The Central Limit Theorem has been used to justify
assuming normality (Gaussian) in an over-wide number of cases. For
a reasonable sample size from a distribution quite different from
the Gaussian, this is a bad assumption. If one knows, or has
reason to believe, that a certain probability function prevails,
then that is the function to use. An argument can be made for not
assuming any "standard" distribution, but using a non-parametric
distribution based on the data itself. This is fine for large
amounts of data and for prediction within the central portion
(say .2 to .8) of the distribution. However, such distributions
are not usually well defined in the tails, especially with small
sample size, so some assumption must be made concerning a distribution function appropriate for these areas. The Logistic
function is often used because of its mathematical tractibility.
For dichotomous (0-1 or pass/fail) data the method of Walker and
Duncan (3) is convenient. Notice, however, that they disregard

physical units in their example of its application. Ignoring the principle of dimensional homogeneity is a dangerous oversight in any model used for extrapolation.

If the data to be fit are continuous there are general non-linear methods which can be used to fit almost any probability function (8), including a variety of so-called probit analyses for (assumed) Gaussian data (9). For many of these methods, convergence is slow or nonexistant if the values initially selected for the fitted parameters are not sufficiently close to the final values.

If the function may be made linear with respect to its unknown parameters by a suitable transformation, then it may be fitted by the Linearized Least Squares method (10) so as to minimize the root mean square error in the original (untransformed) space. The essence of this technique is to use weighted (linear) least squares to effect a non-linear least squares fit. Assume that the equation has been transformed into an equal variance space and let

$$y = \text{the resulting dependent variable}$$

$$\underline{x} = \text{the vector of independent variables}$$

$$\underline{b} = \text{the vector of parameters to be fitted}$$

$$\underline{a} = \text{the vector of known constants}$$

then $\qquad y = f\ (\underline{x},\ \underline{b},\ \underline{a})$ $\hfill$ (1)

The function (1) may be linearized if, through any set of mathematical operations, equation 1 may be transformed into

$$h(y) = \sum_i b_i\ g_i\ (\underline{a},\ \underline{x}) \tag{2}$$

The usual procedure is to employ least squares directly on equation 2. However, this results in minimizing the squared error in h, not y. That is, the procedure finds the set b_i such that the quantity

$$\sum_j (\ \Delta\ h_j)^2 = \sum_j [h_j(y) - \sum_i b_i\ g_i\ (\underline{a},\ \underline{x}_j)]^2 \tag{3}$$

is minimized. What is desired is the minimum of $\sum_j (\Delta y_j)^2$. This may be achieved by iteratively conducting a least squares procedure on equation 2 with weights:

$$w_j^2 = \left(\frac{\Delta y_i}{\Delta h_j}\right)^2 \tag{4}$$

where the Δ's are from the previous iteration. Starting weights are obtained from the differential approximation to the ratio of differences of equation 4; i.e.,

$$w_j^2 = \left|\frac{dh}{dy}\right|_{y_j}^{-2}$$

where the derivative is evaluated at the j th data point to provide the weight appropriate at that point. Unlike most nonlinear methods, therefore, Linearized Least Squares does not require initial guesses, but derives good starting values from the data and the derivative.

A simple example is found in the Logistic function discussed above:

$$y = P = \frac{1}{1+e - (b_0+b_1 \ln x)}$$

In the original space the dependent variable is the probability, P. The equation may be linearized as:

$$h(P) = \ln \left(\frac{P}{1-P} \right) = b_0 + b_1 \ln x = \sum_{i=o}^{1} b_i g_i (X)$$

then: $g_0 (x) = 1$

$$g_1 (x) = \ln x$$

where x is the only independent variable. For the first iteration,

$$\frac{dh}{dP} = \frac{1}{P (1-P)}$$

and $$w_j^2 = \left| \frac{dh}{dP} \right|_{Pj}^{-2} = P_j^2 (1 - P_j)^2$$

we minimize $\sum_j w_j^2 \Delta h_j^2$ which results in the usual weighted least squares, $\min \sum_j w_j^2 \Delta h_j^2 = \min \sum_j w_j^2 (h_j - b_o - b_1 \ln x)^2$.

In their dichotomous fit, Walker and Duncan transform the Logistic function to an equal variance space by dividing each data point by its variance. The variance of a probability value, P, is P (1-P). For the first iteration, P (1-P) is equal to w. This suggests minimizing the function

$$\sum_j \frac{\Delta P_j 2}{w_j} = \sum_j w_j \Delta h_j^2. \quad (5)$$

In linear least squares (unweighted) where $_j \Delta y_j^2$ is minimized, it can be shown that $\sum_j \Delta y_j = 0.$ With weighted least squares, the sum

$$\sum_j \Delta y_j = \sum_j w_j \Delta h_j \quad 0. \quad (6)$$

However, if equation 5 is used (weights w_j rather than w_j^2), then equation 6 does equal zero. When a zero sum of deviations is desirable, function 5 may be minimized, often without increasing the root-mean-square-error by an undue amount.

In conclusion, the following principles may be of some help in modeling in a nonlinear, stochastic universe:

. Model first. Propose as many reasonable models as you

can - then design experiment(s) to discriminate among them.

. For maximum applicability (extrapolation) be consistent with physical laws - including the principle of similitude.

. Whenever none of the proposed models is acceptable, amend the model to fit the data.

Specific to Probability Models:

. Model on means - then fit on all data.

. Normalization is strongly advisable, preferably by dimensional analaysis.

. Transform, if necessary, to equalize variance over domain of definition.

. Stochastic models often require larger data bases than deterministic models.

. Be prepared to seek a nonlinear, stochastic model until it is demonstrated that a linear or deterministic approximation is acceptable.

Literature Cited

1. "Albert Einstein - Hedwig und Max Born: Briefwechsel 1916-1955", Nymphenburger, Munich, 1969.
2. Rosen, R.; _Am J. Physiol_, 1983, 244, R591-R599, "Role of Similarity Principles in Data Extrapolation".
3. Walker, S. and Duncan, D.; _Biometrika 54, 1 and 2_, 1967, 167-179, "Estimation of the Probability of an Event as a Function of Several Independent Variables".
4. Sturdivan, L. M.; "Modeling in Blunt Trauma Research", Second Annual Soft Body Armor Symposium, Miami Beach, FL, Sept 1976.
5. Bridgman, P.; "Dimensional Analysis", Yale University Press, New Haven, CT, 1922.
6. Langhaar, H.; "Dimensional Analysis and Theory of Models", Wiley, NY, 1951.
7. Johnson, N. and Leone, F.; "Statistics and Experimental Design in Engineering and the Physical Sciences", Wiley, NY, Vol II, 1964, 54-56.
8. Marquardt, D.; _J. Soc. Ind. App. Math II_, 1963, 431-441, "An Algorithm for Least Squares Estimation of Nonlinear Parameters".
9. Finney, D.; "Probit Analysis", Cambridge University Press, NY, 1952.
10. Sturdivan, L. M. and Jameson, J.; "Linearized Least Squares", _Proceedings of the 1976 Army Numerical Analysis and Computer Conference._ ARO Report 76-3, US Army Research Office, 1976.

RECEIVED August 6, 1984

INDEXES

Author Index

Subject Index

Jacket design by Pamela Lewis

Elements typeset by Hot Type Ltd., Washington, D.C.
Printed and bound by Maple Press Co., York, Pa.